90s BITCH

MEDIA, CULTURE, AND THE FAILED PROMISE OF GENDER EQUALITY

ALLISON YARROW

HARPER  PERENNIAL

NEW YORK • LONDON • TORONTO • SYDNEY • NEW DELHI • AUCKLAND

HARPER ● PERENNIAL

HarperCollins books may be purchased for educational, business, or sales promotional use. For information, please email the Special Markets Department at SPsales@harpercollins.com.

FIRST EDITION

Designed by Leydiana Rodriguez

Library of Congress Cataloging-in-Publication Data has been applied for.

ISBN 978-0-06-241234-8 (pbk.)

HB 09.12.2023

For Ben, Ruby, and Oscar

CONTENTS

PROLOGUE

Many women remember the first time they were called a bitch in pristine detail, like a first kiss or childbirth. For me, it was at a party for my high school soccer team where I got drunk for the first time. An argument with a friend about a boy escalated into yelling and she called me a bitch. I was so startled that I slapped her in the face. It was the talk of the lunchroom the next week, in part because we had just learned about irony in English class and my friend's last name happened to be Slappey. Being the perpetrator was humiliating. Girls didn't hit and I had violated the code. But I had to retaliate because innately I knew that being called a bitch was the worst possible slight.

"Bitch" is a gendered insult with a long history of reducing women to their sexual function. Ancient Greeks slandered women by calling them dogs in heat who begged for men—a slur that referenced the virgin goddess Artemis, the huntress who changed herself into a wild dog. According to etymologists, the word has long been used with the intent of "suppressing images of women as powerful and divine and equating them with sexually depraved beasts." From its very conception, "bitch" was a verbal weapon designed to restrain women and strip them of their power.

Today, "bitch" has been spit-shined, retooled, and given new life. We hear women using it to describe one another—"boss bitch," "basic bitch," and "resting bitch face" are ubiquitous terms on social media, in the school lunchroom, and around the office watercooler. What was once a derogation is now seen as an appellation of empowerment and sisterhood. But the attempted reclamation of the word doesn't change its history or more common use: it has historically been, and remains, the worst invective hurled at women—one that degrades, disparages, and disenfranchises all at once.

This is plainly on display in the historical record. Use of the word has increased as women have gained power and influence, specifically to undercut their achievements and stop their progress. Indeed, this is the real story of how "bitch" and its corollaries were deployed by misogynists in the 90s, and how the word and the concept proliferated throughout society in that decade. This "bitch bias" shaped the way a generation of women and men came of age, and also this current moment. We can no longer ignore the history of "bitch" and how it has influenced the world we live in today.

I'll use the verb "bitchify" and the noun "bitchification" to characterize how 90s media and societal narratives reduced women to their sexual function in order to thwart their progress.

INTRODUCTION

'm not sure whether to follow the girl in the Hanson tee or the guy in the *All That* hat. They are walking in opposite directions. If they are both headed to the inaugural 90s Fest—a Nickelodeon-sponsored outdoor concert featuring a scattershot assemblage of popular bands from that decade—then somebody is lost.

The large lot on the East River in Williamsburg, Brooklyn, features a slime machine opposite the large stage. Contest winners, Pauly Shore, and the rap duo Salt-N-Pepa will later be doused in green goo. Life-size Jenga and Connect Four draw a few players. Girls are sprawled atop a leopard coverlet in a "Real-Life 90s Girl Bedroom" sponsored by Shop Betches.

The 90s may be the frame, but this is still a 2015 music festival. The eight-dollar hot dogs are named for performers, and the wristband handlers are high. Attendees wear the decade's full regalia—baby tees and butterfly clips, combat boots, flannel, acid-washed jeans, oversize blazers, leggings, kinderwhore, neon. Some are dressed as the Spice Girls. Others wear Clinton/Gore 1992 T-shirts that look conspicuously white and crisp for twenty-three years later.

Children of the 90s are a demographic relatively new to the workforce and to their own money, and businesses want to lock them down for life. Their childhood television programmer, Nickelodeon, wants them back, too. The network is promoting a new block of programming called the Splat that will air the shows attendees watched as kids.

I am one of these 90s kids. I was eight years old at the start of the decade and eighteen at its end. It's easy to rhapsodize about the years spent shedding childhood, and I have warm memories of mine. I collected the stuff—the American Girl doll, the baby tees, the Lisa Frank Trapper Keeper—and mainlined the culture. I loved films like *Clueless* and *Reality Bites* and devoured book series like The Baby-Sitters Club and Sweet Valley High. I watched tabloid talk shows and MTV, and learned to drive listening to Nirvana and Lauryn Hill on compact discs.

The nostalgia strategy Nickelodeon is banking on seems inspired by how we hear the music of our youth. Brain-imaging studies reveal that the deep attachment we feel to the music from our adolescence isn't a conscious preference or reflection of critical listening, but the result of a host of pleasure chemicals bombarding our brains. Despite our tastes maturing, 1990s daughters and sons will likely prefer TLC, Smash Mouth, and New Kids on the Block to new hits—not for their quality, but for their emotional wallop. Perhaps that's one reason why clubs from Brooklyn to Portland have found success with 90s Nights, drawing thousands to reminisce and dance to the songs they once loved.

This onslaught of 90s nostalgia is no great surprise, as kids of the 90s tumble into adulthood, bidding reluctant farewell to their younger zine-reading, Game Boy–playing, *Rugrats*-watching selves. They are pondering having children of their own, or are newly minted parents. Nostalgia is a gift and affliction of every generation. It eases the collective identity crisis as adulthood's mundanities gel.

It's also unsurprising that 90s Fest presents a version of the 90s that doesn't attempt to deviate much from history's script. And the children of the 90s don't seem to want it to. Reheating and serving the commercial culture that we 90s kids remember is just fine, thank you. In between band sets, a jumbotron plays a montage of clips from 90s television shows, movies, music videos, and advertisements. Festival goers intermittently sigh and aww at the *Saved by the Bell* credits, Sunny Delight ads, and Freddie Prinze Jr. But the pull of the past has clouded our critical minds. Are the 90s really as great as we remember them to be?

THE HIGHLY ANTICIPATED DECADE OF WOMEN

As the decade dawned things were looking up for women. Daughters of second-wave feminism came of age and chose new paths unavailable to their mothers: delaying marriage and children, pursuing higher education, joining the workforce, and assuming independence and identities outside of the home. The gaps between men and women in education "have essentially disappeared for the younger generation," declared a 1995 report by the National Center for Education Statistics. At that time, female high schoolers bested their male counterparts in reading and writing, took more academic credits, and were more likely to go to college. By 1992, they earned more bachelor's, master's, and associate's degrees than men. The equal education promise of Title IX was coming to fruition.

In the 80s, women began marrying older, or not at all. For

more than a century, the median marriage age for women swung between twenty and twenty-two, but in 1990, it nearly jumped to twenty-four. By 1997 it reached twenty-five. Carefree sex outside of marriage became increasingly acceptable. Access to birth control expanded. Postponing marriage and kids liberated women sexually; it also gave them increased economic power and paved their entry into male-dominated careers. By the decade's end, women accounted for close to 30 percent of lawyers, nearly half of managers, and more than 40 percent of tenure-track professors. Almost half of married women surveyed in 1995 reported earning half or more of their total family income, leading the study's sponsor to declare, "Women are the new providers."

The forward motion of the 90s seemed to build on the 80s, a decade of hallowed female pioneers in diverse fields. Sally Ride traveled to space. Geraldine Ferraro secured the vice presidential nomination of a major political party. Alice Walker and Toni Morrison won Pulitzer Prizes for their epic, women-centered fiction. Madonna smashed barriers in music, entertainment, and popular culture. Because these firsts and many others were so widely celebrated, society assumed these trailblazing women would also cut a path for all women to advance in work, entertainment, politics, and culture in the years to come. At last, the dream of gender equality would be realized.

The dream, as we know, was not realized. But a quick glance back at the 90s would suggest that American women indeed made significant progress during the decade. In Janet Reno, Madeleine Albright, Judith Rodin, and Carly Fiorina, the 90s saw the first woman attorney general, secretary of state, president of an Ivy League institution (University of Pennsylvania), and CEO of a Fortune 100 company (Hewlett-Packard). More women won political office than ever before in 1992, the so-called Year of the Woman, when their numbers in the Senate tripled (from a measly two to a small but more respectable six).

Cultural feminism in the 90s made strides, as well. The "Girl Power" movement promised that progress for women would trickle down to girls, too. Indie subcultures defined by girl-made zines, music, art, and websites flourished, providing young women new platforms for self-expression. Girl culture was reclaimed and celebrated by the Riot Grrrl movement, *Sassy* magazine, websites like gURL.com, government initiatives, subversive feminist musicians, and independent films.

Indeed, the 90s was a decade in which women were front and center—but not in the way we like to remember.

"'BITCH' IS LIKE A TITLE. I'VE SEEN THEM OWN IT."

Backstage at the 90s Fest, I ask Coolio why rappers in the 90s were so enamored with the word "bitch." He should know. His song "Ugly Bitches" excoriates them as lazy, slutty, and deserving of murder.

"They were a bunch of nerds. They didn't have no game. A lot of rappers acted like they hated women," he says, excluding himself.

Why? I ask. Could they have done better by women?

"Rappers didn't do anything to women. You know how many women got rich off rappers? How many women rappers made rich? You got to take the good with the bad. He might call you a bitch, but most women say, 'Well, I'll be a bitch if I got three zeros or six zeros on my bank account. I'll be two bitches,'" he replies. "'Bitch' is like a title. I've seen them own it."

Coolio wasn't the first to deploy this particular articulation of "bitch." "Bitch" was defined in 1811, in *Slang and Its Analogues, Past and Present*, as "a she dog" and "the worst name one could call an Englishwoman, more provocative and insulting than 'whore.'" A whore was at least paid for sex, but a bitch gave it away for free. Being sexually needy while lacking capitalistic acumen made a bitch all the more detestable. Nearly two hundred years later, this meaning found its way into contemporary music.

Hip-hop was largely created by two groups who, to craft masculine, flinty personas, separated themselves from women by deriding them as bitches: Public Enemy and N.W.A. Public Enemy's 1987 hit "Sophisticated Bitch" describes a "stone cold freak" bedding "execs with checks" and "boys from the dorms" in the same week. This bitch incites violence. "People wonder why did he beat the bitch down till she almost died?" the group sings.

Public Enemy needed to appear hard. Chuck D, Flavor Flav, and crew pioneered hip-hop after meeting at the liberal arts school Adelphi University on Long Island amid suburbs and swimming pools, not the gunshots and hard streets they rap about. Through that lens, "Sophisticated Bitch" looks like a vehicle to diminish women, to achieve the steely, gangsta rep that a childhood of cul-de-sacs and BBQs couldn't provide.

N.W.A, comparable to Public Enemy in their contributions to hip-hop, put out their own bitch anthem in 1987, "A Bitch Iz a Bitch," implicating their entire female audience. "Now ask yourself, are they talking about you? Are you that funky, dirty, money-hungry, scandalous, stuck-up, hairpiece contact wearing bitch? Yep, you probably are . . . Bitch, eat shit and die (ha ha)."

As in physics, in popular culture, every action has an equal and opposite reaction. In the 90s, against the backdrop of bitchification normalized by N.W.A and Public Enemy, among others, feminists attempted to reclaim the word for their own purposes. The reclamation spread through media. *Bitch* magazine promoted feminism to young women, inspired by gays and lesbians who had successfully taken back "queer" from its oppressive uses. The term appeared on novelty items like T-shirts and buttons. Countless products marketed in the 90s celebrated the bad-girl sex icon—the bitch—and accorded her power. Elizabeth Wurtzel published *Bitch: In Praise of Difficult Women*. Her definition included being able to "throw tantrums in Bloomingdale's." Pop icon Madonna embraced it, saying in an interview, "I'm tough, I'm ambitious, and I know exactly what I want. If that makes me a bitch, OK." Singer Meredith Brooks crooned that she was one—along with a lover, child, mother, and host of other female identifiers—in her hit song, "Bitch." Its incessant radio play in the late 90s suggested that, for women, being a bitch was not only kosher but aspirational.

By the end of the decade, however, the promise of equality for women was revealed to be something between a false hope and a cruel hoax. Parity, it turned out, was paradox: The more women assumed power, the more power was taken from them through a noxious popular culture that celebrated outright hostility toward women and commercialized their sexuality and insecurity. Feminist movements were co-opted. Soon, women would author their own sexual objectification.

Many 90s efforts to take back "bitch" were rooted in consumerism. "Bitch" and its villainous corollaries became a bad-girl identity to sell—a sly marketing tool. The advertisers' bitch was a sexy villainess at the mall and in Hollywood. When Chanel debuted a red-black nail polish shade, Vamp, in 1994, stores fought to keep it in stock. Other makeup purveyors released nail and lip colors with names like Vixen, Wicked, and Fatale. But beneath the message that buying bitch gear was empowering lurked a sinister trait.

In the 90s, the music, media, and products freely portrayed women as bitches and every nasty offshoot imaginable. The generation's youth, and girls in particular, internalized this term and saw its power to offend and undercut even the most powerful women—the First Lady, the secretary of state, and the attorney general, to name a few. I believed such descriptions of public women and didn't question them, and I know I'm not alone. Anita Hill lied, Monica Lewinsky was a tramp, Marcia Clark was unqualified, and girls were supposed to "go wild" for cameras or be nasty because it was liberating and empowering. The public colluded in degrading women, and an entire generation of girls grew up in the fog of a bitch epidemic. Our would-be role models were pilloried, our authentic selves were pried away from us, and to get them back we were told to buy lingerie and magazines featuring sex tips.

THE DECADE THAT DESTROYED WOMEN AND POISONED GIRLHOOD

In the end, the 1990s didn't advance women and girls; rather, the decade was marked by a shocking, accelerating effort to subordinate them. As women gained power, or simply showed up in public, society pushed back by reducing them to gruesome sexual fantasies and misogynistic stereotypes. Women's careers, clothes, bodies, and families were skewered. Nothing was off-limits. The trailblazing women of the 90s were excoriated by a deeply sexist society. That's why we remember them as bitches, not victims of sexism.

The 90s bitch bias is so pervasive, so woven through every aspect of the 90s narrative, that it can actually be tough to spot. Stories of notable women in the 90s almost invariably suggest they were sluts, whores, trash, prudes, "erotomaniacs," syco-phants, idiots, frauds, emasculators, nutcrackers, dykes, and succubi. These disparagements were so embedded in the cul-tural dialogue about women that many of us have never stopped to question them. I spoke with more than a hundred women about their remembrances of the 90s, and the majority of them internalized 90s bitchification, too. The stories of 90s women have become sexist mythology, an erroneous history that saps women of their power, just as it was intended to do. Indeed, the aftershocks of 90s bitchification ripple into contemporary society. Discrediting women based solely on their gender, sex-ually harassing them, and reducing them to their fuckability endures today from the school yard to the boardroom in part because this was, writ large, ubiquitous and accepted behavior in the 90s.

I loved my 90s childhood. But it wasn't until returning to this decade as an adult that I came to see how mainstream 90s narratives in media and society promoted sexism and exploited girlhood. I wrote this book because I was utterly shocked by what I found while investigating 90s narratives about women. The decade is barely considered history. It was supposed to be the modern era, with doors flung open to unprecedented advancement for women and gender equality. But 90s bitch-ification was like water flowing into every crevice. It existed everywhere I looked, which is why this text is by no means exhaustive. The stories I've chosen to reexamine are the ones that I believe reached the furthest and have had the most resonance. Taken together, they explain the status of women in American society today.

1

PRETTY ON THE OUTSIDE

It's no coincidence that 90s bitchification coincided with a radical new media landscape. The emerging twenty-four-hour news cycle—providing real-time, unremitting coverage of live and current events—swiftly infiltrated households and shaped the American consciousness during and after the Persian Gulf War. When the US military bombed Iraq in January 1991, television cameras followed, and the first twenty-four-hour cable news network, CNN, flourished. Americans didn't just watch—they binged. More than half claimed they were addicted to watching the war on TV, according to a 1991 Times-Mirror survey. A majority of adults under thirty dubbed themselves "war news addicts," and 21 percent admitted that watching disrupted their focus on their jobs and normal lives.

When the war concluded mere weeks later, CNN had the round-the-clock infrastructure and an insatiable audience; all they needed was another war. But soon they would learn that political dramas, crime, and Hollywood were far cheaper to cover, and often more popular, than bombs over Baghdad.

This continuous, addictive format produced an unrelenting fixation on public figures and news makers, but none so much as the women gaining power and prominence in the 90s. Women touched by scandal, whether they were alleged perpetrators or victims, were hounded by the press. When any woman made the news, she often stayed there for days, weeks, months, and, in some cases, years. Meanwhile, news consumers blamed women for their own unceasing visibility, as if they had narcissistically engineered unflattering coverage of themselves for personal gain.

Another consequence of the rise of the twenty-four-hour news cycle was that the nation consumed the very same stories at the very same time. During the 90s, some of us read newspapers and magazines, listened to the radio, and began consuming news online. But television was by far the largest media stage of all. Many of the decade's biggest news stories centered on women as TV news increasingly became episodic infotainment. The compulsive focus on scandal-driven tales pitching women out front and center, or against one another, gave new meaning to the old adage "If it bleeds, it leads."

Women didn't fare any better on fictional television. Television's ideal woman in the late 80s and early 90s was "beautiful, dependent, helpless, passive, concerned with interpersonal relations, warm and valued for her appearance more than for her capabilities and competencies," according to a 1992 book-length report by an American Psychological Association task force. This TV dream woman was no real woman at all, but a degraded caricature of one. After analyzing five years of television programming, the APA authors concluded that TV completely ignored its most devout audiences—women, minorities, and the elderly. The proportion of women characters appearing on primetime television barely budged throughout the 90s, moving from 38 percent in the 1990–1991 season to 39 percent in 1998–1999—a mere one percentage point increase in nearly a decade.

The way women appeared on television in the 90s is easily traced to their underrepresentation both in front of the camera and behind it. Martha Lauzen, a San Diego State University professor, found that only 15 percent of the creators of the top one hundred primetime shows in the 1998–1999 season were women. Women were only 24 percent of executive producers, 31 percent of producers, 21 percent of writers, 16 percent of editors, and just 3 percent of directors. Television misrepresented and objectified women, while its staffers were mostly male. Women weren't telling their own stories; men were telling the stories of women that they wanted to see.

If 90s television was ground zero for the war on women, the soldier on the front line was *Beverly Hills, 90210.* The father of jiggle television himself, *Charlie's Angels* producer Aaron Spelling, applied said jiggle to high school with his *90210.* In the series, young beauties in revealing clothing luxuriate in wealth, among palm trees, and on the perpetual brink of boinking one another. Its objectification game was strong. The program's credits open by panning up a bikinied body, but the shot cuts away before landing on her face. The 1991 season two premiere was viewed in close to eleven million households, and the show became one of Fox's top performers. The following year, half of teenage girls polled named it their favorite show.

90210 is set on the lush campus of West Beverly Hills

High School, which resembles a fancy college. The main male characters—Brandon, Dylan, and Steve—talk about their emotions and wear wounded looks and sexy sideburns that my grade school friends and I daydreamed about. They are independent and active: playing sports, writing for the school paper, working jobs, and even living alone. Susan Douglas's book about misogyny in modern pop culture, *The Rise of Enlightened Sexism*, points out that while the male characters contend with the show's meatier dilemmas, the female protagonists—Kelly, Brenda, and Donna—are relegated to obsessing over shopping, gossip, and dating cute boys. Their power is concentrated in their looks, how they fill out bikinis, and their likelihood to have—or be victimized by—sex. Women succeed in the show in proportion to their sexiness.

In the bubble of *90210*, girls' bodies exist for enjoyment and ridicule. In the pilot episode, Steve taunts Kelly mercilessly about her summer nose job and asks her, "What's next, tummy tuck? Liposuction?" as he ogles her backside. Kelly believes the surgery has made her pretty and popular. The other girls follow her lead. "Brenda tries to dye her hair blonde in order to impress Dylan, but winds up looking like a clown" is a logline of one early episode. After making out with Dylan for the first time, Brenda trades her wardrobe of buttoned-up shirts for plunging V-necks and midriff-baring tops. One female character who rejects this path is the smart (and poor, and Jewish) newspaper editor, Andrea Zuckerman. She challenges boys' ideas, bosses reporters, and wears baggy suits. Her punishment is that other characters treat her like a nuisance, and she is incurably single for much of the show.

Sex, for the girls, careens wildly between dangerous and

frivolous. They are graded for performance and blamed when things go wrong. Steve tells David he "dumped" Kelly because she's "got a nasty personality" and is "lousy in bed." "I could live with that," says David. Steve bullies Kelly in hopes that she'll return to him, which she does, sleeping with him one night at a party. This troubling story line exalts harassment as a means to an end: Kelly is just a tease, and girls like her will give it up eventually.

Brenda's journey to lose her virginity to Dylan is a cultural touchstone for many 90s kids. Her preparation for the occasion is resplendent with a nearly four-minute montage of Kelly and Donna dressing her for it. The couple has already discussed the act and its potential consequences. What should be empowering—Brenda makes an informed choice that she is happy about—becomes shameful in a later episode, after livid feedback from young viewers' parents who disapproved of her confidently choosing premarital sex. Brenda feels good about losing her virginity, but later reneges. "I love Dylan and I thought I knew what I was doing, but I'm beginning to get the feeling that it wasn't worth it," she says.

The girls endure more body shame throughout the series. Brenda has unplanned-pregnancy and breast-cancer scares, and blames herself for these false alarms. Male characters period-shame the girls, and the girls period-shame one another. "Can't you stop thinking about guys for one second? There's more to life," Brenda snaps at Kelly, who retorts, "Sounds like it's that time of the month."

Sex on *90210* is portrayed as transactional, with young women's bodies traded as currency. It's no surprise, then, that beauty brands popular with young customers launched *90210* tie-ins. Softlips, Noxzema, Conair, and makeup-organizer Sass-aby were among the companies that sold some $200 million in *90210*-branded products between 1990 and 1992. *90210* trading cards appeared in Honey Nut Cheerios boxes—a cereal adver-tised to very young children. One marketing firm found kids as young as seven liked the show. A youth marketing executive told the *Los Angeles Times*, "The younger kids think it's really cool. It makes them feel like teens." A thirteen-year-old Manhattan pri-vate schooler revealed to the *New York Times* that his interest in sex began after watching *90210*. "The people were cool. I wanted to try what they were doing on the show," he said.

FAT GIRL SLIM

If *90210* schooled a generation of girls that their bodies were for sex, a recurring 90s television character reinforced that trou-bling lesson: the fat girl who did good and got sexy and thin. Monica Geller on *Friends* (Courteney Cox) and Helen Chapel on *Wings* (Crystal Bernard) were both repeatedly identified and ridi-culed as formerly fat. On both shows the characters are played by beautiful starlets with arms the size of toilet paper tubes, making it hard to believe that they were ever anything but lithe. Yet their characters' former large-girl status is a source of humor on both programs and drives much of their pathos.

Monica and Helen must pay for their prior fatness, which will always define them. Other characters never seem to let them live it down. On *Friends*, fat jokes flow like cappuccinos at Cen-tral Perk. "Some girl ate Monica," Joey says when the gang finds a video of her larger self. Monica's former fleshiness is played as a recurring fat-shaming laugh line throughout the series. In one episode, the painfully skinny Courteney Cox dons a fat suit and wiggles around in compromising poses—holding a doughnut, for example—for laughs. She orders a pizza for a college party, then hollers when it arrives and eats it in front of everybody. Monica is her most uninhibited self when she wears the fat suit. She's actually a lot more fun. But "fat" Monica and "fat" Helen are fantasies in the worlds of these shows; the real characters are neurotic skinny women with accentuated collarbones. They're so happy to no longer be fat that they don't even defend themselves against jokes about their former weight.

Apart from providing a humorous subplot in both shows, Monica's and Helen's off-camera weight loss is fetishized, permitting other characters to sexualize them. Losing her girth allows Helen to become a sexual prize to the show's leading men, the Hackett brothers, Joe and Brian. "You've lost weight," one says to her. "Maybe a pound or two. Or sixty!" she replies, for big canned laughs. When Joe tells Brian that he can't date Helen because she's "like a baby sister," Brian responds, creepily, "I think it's time I gave that little tyke a bath."

Chandler mocks Monica's size in front of her family one Thanksgiving, only to want to sleep with her the next, after she sheds weight. "Oh my God. You look so different! Terrific. That dress. That body," Chandler says, leering at her. They end up marrying.

Plotlines like these perpetuated a societal mandate of thinness, prompting girls to seek skinniness to be loved, just like the television characters. Hundreds of pages of advertising and editorial content in magazines like *Teen, Seventeen,* and *YM* contained examples of the body ideals that compelled girls to want to lose weight. Nearly 70 percent of elementary school girls said that magazine pictures affected their conception of a body ideal, while almost half said they made them want to lose weight. I remember the wordy, vague weight-loss ads and advertorials in these glossies, encouraging girls to send away $12.95 (plus $3.00 shipping and handling) to sketchy programs with names like Special Teen Diet and the Clinic-30 Program. Black-and-white advertisements featured happy-looking girls in bathing suits or diary confessionals seemingly written by teenage girls thrilled with their diets' results.

I bought these magazines and wondered if the testimonials were real. I never sent away for these scams growing up, but by the time I was eleven I did consider it. The CDC reported in 1995 that girls were nearly twice as likely as boys to believe they were overweight. It's no wonder Riot Grrrls took to drugstores, newsstands, and 7-Elevens in the 90s, armed with scraps of paper to covertly slip between these magazines' pages urging, "You don't need lipstick to be beautiful," "This magazine wants you to hate your body," and "Love your body the way it is."

Popular touchstones like teen magazines and *Beverly Hills, 90210* normalized the commodification of the female body, leading girls to believe that the quest for self-worth worked best from the outside in. The television show's scads of marketing tie-ins were often the same as magazines' advertisers—makeup, acne treatments, shoes, perfume, and denim—reinforcing the idea that with enough money, girls could buy body insecurity away. This concept was the target of Naomi Wolf's 1991 book, *The Beauty Myth*, which investigated how women are saturated with unrealistic images of perfection and hoodwinked into thinking that their value is found solely in their exteriors, leading them to buy numerous fixes. But this was a book for women, and these realizations often eluded girls.

Naturally, the 90s saw an uptick in the incidence of eating disorder diagnoses and discussions about anorexia and bulimia, leading experts to declare an epidemic. The most alarming discovery about disordered eating and negative body image was how shockingly early the signs began. In 1991, more than 40 percent of first through third graders wanted to be thinner. Nearly half of nine- to eleven-year-olds admitted to dieting "sometimes" or "very often," while between 40 and 60 percent of adolescent girls said they had dieted, fasted, made themselves vomit, or taken laxatives or diet pills. Since 1995, the percentage of girls who believe that they are overweight has tripled.

It goes without saying that eating disorders disproportionately afflicted women. But body anxiety was making women sick in more ways than one. Cultural aspiration to thinness could explain why twice as many women suffered from depression as men, according a 1990 paper by Mandy McCarthy, a professor of psychology at the University of Pennsylvania. In "The Thin Ideal, Depression and Eating Disorders in Women," she argued that body dissatisfaction coupled with societal mandates of thinness is more likely to cause depression in women than in men.

All the while, celebrities continued to shrink. *People* magazine documented the "waif wave" trend in 1993, exemplified by Kate Moss, who looked as if "a strong blast from a blow dryer would waft her away." Moss certainly didn't cause anorexia. But many anorexics cited her picture "as an ideal," according to an eating disorders counselor who told the magazine, "I haven't seen that with any other particular model before."

The tabloid press wondered whether the entire female cast of the hit show *Ally McBeal* had launched a starvation competition. Critics attacked Calista Flockhart, Courtney Thorne-Smith, and Portia de Rossi for their thinness, as their bodies waned with each successive season, but the actresses didn't admit that they were practicing extreme dieting and exercise. The takeaway for women watching the show, or seeing the shrinking frames of these celebrities in magazines, was either that they were naturally that thin, or that they were somehow better at achieving the lithe body ideal than the masses. Women looked at thinning female bodies like these and hated their own.

It wasn't until much later that Thorne-Smith and de Rossi opened up about their diseases in the press. Thorne-Smith gave interviews admitting to unhealthy diet and exercise regimens while working on the show. De Rossi published the memoir *Unbearable Lightness* in 2010, chronicling the disease that left her at eighty-two pounds and approaching organ failure. But in the 90s, widespread celebrity thinness seemed to deny that eating disorders and extreme dieting were problems.

POWER PANTIES

Meanwhile, as young women carved themselves into more perfect pieces and fretted about the judgments and measurements of their bodies, something was happening to their underclothes. By the 90s, underwear no longer toiled as mere scaffolding for women's wear. As Madonna foretold in 1990, when she stormed onto television in her cone bra, fashion lingerie became the main event and "assumed a new presence in the lives of Americans," according to the *New York Times*. Marketers sold women the bromide that power was a pair of tiny panties. They simultaneously promised men that purchasing panties for their lovers was the key to controlling the sexual experience.

For most women, who wore Jockey and Hanes basics in white and nude, or black for special occasions, Victoria's Secret was a revelation. The growth of the lingerie retailer, and the expansion of sexy lingerie to suburban women, can be attributed to a male founder's vision. Roy Raymond was so uncomfortable buying underwear for his wife, Gaye, that he started a business to ease the burden. Surely other red-blooded men could relate. "When I tried to buy lingerie for my wife," Raymond told *Newsweek* in 1981, "I was faced with racks of terry-cloth robes and ugly floral-print nightgowns, and I always had the feeling the department-store saleswoman thought I was an unwelcome intruder." In other words, since its inception, Victoria's Secret's raison d'être has been catering to men.

By 1993, Roy and Gaye's small Palo Alto shop—modeled after and named for their Victorian home—had grown to nearly six hundred stores, riding the crest of the great suburban mall invasion of the 90s. Shopping malls allowed Victoria's Secret to supplant the country's obsession with political correctness and family values with an alluring S&M vocabulary. Suddenly housewives and coeds were talking about bustiers, garters, teddies, and merry widows—once-obscure garments that could now be purchased in the same trip to buy kitchenware or pick up a cookie cake. What had once been the purview of pornography, strip clubs, sex dungeons, and Hollywood was now just a car ride away in suburbs of the North Shore or Short Hills. The *New York Times* credited Victoria's Secret with helping turn "what was once a discreet (or salacious) business into the fastest growing segment of the nation's $6.4 billion women's intimate apparel industry."

Another reason mass-market undergarments morphed from support to foreplay was the booming mail-order catalogue business. Why drive to a store when you could just walk to your mailbox? By 1992, more than $51.5 billion was spent on mail-order goods, and half of all adults bought merchandise from catalogues.

Lingerie catalogues offered colorful spreads of nearly naked women posing among the trappings of wealth—Oriental rugs, tasseled curtains, and crystal chandeliers—to create desire. The models themselves were emissaries of sex, and women could aspire to the kind of pleasure presented in the pages and shop in complete privacy, without ever leaving their homes. Men could stash the catalogues in their bathrooms next to *Playboy* and the *Sports Illustrated* Swimsuit Issue. By 1997, Victoria's Secret was shipping out 450 million catalogues a year and banked $661 million in mail-order sales alone. Catalogues concocted a whole fantasy world that begged women to enter and find their beautiful, newly sexually confident selves. Victoria's Secret telegraphed that empowerment through pleasure was available to purchase in sizes A to DD.

Where Victoria's Secret differed from its forebears was its models. While most lingerie models—like those in the Frederick's of Hollywood catalogue—seemed fresh off hardcore pornography film sets, Victoria's Secret instead cast sexy girls-next-door. Their wholesome appeal suggested that you could achieve their look if you simply shelled out for a bra or a teddy. Victoria's Secret catalogue models in the 90s weren't the fembot blondes covering *Playboy*, like Anna Nicole Smith and Pamela Anderson (who modeled for Frederick's), but "good girls, well brought up, slightly nymphomaniacal, but only behind closed doors and only when they're in love," explained journalist Holly Brubach in the *New York Times Magazine* in 1993.

At the same time, Victoria's Secret's brand pushed the shopping experience as a sisterhood, "unquestionably a female enclave, like a beauty salon or a harem—a place to which women retreat to make themselves more attractive," Brubach wrote. The prime Victoria's Secret patron was the "Cosmopolitan Girl," a single, career-oriented woman who had sex for pleasure. She could deploy disposable income to perfect the garments beneath her power suit. The creation of the fashion lingerie market, spearheaded by Victoria's Secret, was a natural extension of women remaining single longer and creating their own wealth. The message in the selection, attitude, and staging of Victoria's models told women that purchasing a push-up bra could actualize their own fantasies alongside men's, securing not just desirability, but love, as well. This seemed like new power.

The brand exploded because it succeeded in marketing to women what men wanted. The authority of male desire is coded in the lingerie giant's DNA. Victoria's Secret attracted male customers by prioritizing their need to feel comfortable with underwire, and giving them permission to direct the sexual experience as Raymond had hoped. At Victoria's Secret shops and in catalogues, men's "presence and power . . . are implied in the women's diligent efforts to please them," Brubach wrote. By 1993, Victoria's Secret had reached $1 billion in revenue.

THAT THONG TH-THONG THONG THONG

One particular cut of underwear propelled lingerie sales in the 90s. The thong became so popular that one report likened its gain on the panty to the unseating of ketchup by salsa as America's favorite condiment.

Some say thong fever began with Monica Lewinsky. After she lured the president by flashing hers—earning her the sobriquet "thong snapper"—the undergarment took on a cultural life of its own. When news of Lewinsky's thong incident broke in 1998, the undies were no longer a marker of sleaze, relegated to strip clubs and Frederick's of Hollywood catalogues. By 1998, fourteen million thongs were sold, representing some 40 percent of total panty sales. Thongs became "the fastest growing segment of the $2 billion a year women's panty business," reported the *Wall Street Journal* in 1999.

"You lifted the back of your jacket and showed the president of the United States your thong underwear," Barbara Walters chided in an interview with the disgraced intern. "Where did you get the nerve? I mean, who does that?" Lewinsky laughed to defuse the slight. "That was how our flirtation relationship was progressing," she answered. "It was a very—I know that it sort of has been highlighted. I'm not going to demonstrate for you. But if you take my word for it, it was a small, subtle flirtatious gesture. And that's me," Lewinsky said.

"Was it saying, 'I'm available'?" Walters followed up.

"Well, I think it was saying, 'I'm interested, too. I'll play,'" Lewinsky said. The message was received. She was invited to see the president in private for the first time that night. The thong became "the garment that shook a presidency."

Female sexual ability was telegraphed by the thong's presence alone. The humble underwear had gone from functional to powerful. What was designed to be hidden suddenly demanded to be seen. Thongs, looking like "whale tails," began to peek out of low-rise pants. Celebrities from Halle Berry to Christina Aguilera displayed their G-strings. This was popular not only on red carpets and in clubs, but also in high school hallways. Clothing became tighter and sheerer to contend with the popularity of wearing thongs. Women claimed to purge their closets of traditional panties to make room for thongs and thong-based wardrobes. Suddenly, it wasn't just about wearing a thong, but about letting it be known you were wearing one, too. Underwear as outerwear projected sexual availability and prowess.

While Victoria's Secret profited from the thong trend it abetted, legions of girls who came of age during the 90s were sold a bill of goods. We were taught that sexual fulfillment and love are derived from consumerism—stringy, lacy purchases wrapped in pink heart paper. But thongs, of course, are not a particularly effective means of achieving self-worth. And more worryingly, their proliferation in 90s culture normalized a male-engineered brand of performative sexuality that didn't necessarily translate into happiness or satisfaction for women. In fact, it reinforced the fallacy, pervasive in 90s media and marketing, that women should design their appearance and conform the expression of their sexuality to fulfill male desire.

BIONIC, BREASTY, AND BLONDE

The model body type that sold thongs and fashion lingerie was mainstreamed by a cadre of entertainers, most notable among them Pamela Anderson and Anna Nicole Smith. With their perfectly spherical breasts, hairless limbs, fat-lipped pouts, and eyebrows that looked drawn with a protractor, Anderson and Smith appeared in television, film, and print as glistening envoys of sex. Their look signaled a new normal that women were encouraged to emulate—the "human Barbie doll."

They both shot to stardom through the Playboy Mansion. Smith became *Playboy* magazine's Playmate of the Year in 1993, and Anderson would cover fourteen editions, including the magazine's final nude issue in 2016. Their roles on television and in film were limited to variations of sexpot. Anderson, wearing tiny shorts, trotted out power drills and water heaters as Tim Allen's handy assistant on *Home Improvement*. She graduated to the perfect slow-motion jiggle on the beaches of *Baywatch*. The show's reach at its peak was staggering—more than two billion people each week in 104 countries and on every continent save Antarctica watched Pam run.

Throughout the 90s, these entertainers starred in marketing campaigns that sold sex alongside products. Anna Nicole Smith became the face (and ass) of Guess Jeans, and later the diet supplement TrimSpa. Anderson appeared in beer campaigns, but her DIY sex tape, shot on her honeymoon with Mötley Crüe drummer Tommy Lee, was her most marketable moment of the 90s. The tape was reportedly not intended for public consumption, but rather stolen by an electrician from their home in 1995 and later shared online. It furthered Anderson's reputation as both "sizzling sweet eye candy" and an "out-there sex weirdo." Not long after the tape's release, its content was available on skin channels at nice hotels for around ten dollars.

There's something campy and charming about it, as if for a moment you could imagine that they are not celebrities, but just regular folks doing it on a johnboat. Anderson even asks Lee on the tape, "When are you going to get me prego?" But they were freakishly beautiful and on a yacht. It was the first celebrity sex tape occurring at the nexus of tabloid culture and the internet. The tape birthed a genre that would later anoint stars like Paris Hilton and Kim Kardashian.

Throughout the 90s, Anderson and Smith were ogled by interviewers and queried about their breasts, which they paraded in small outfits. Anderson referred to hers as "props," assisting the creation of her character. The *Chicago Tribune* called them "erotic cartoons." Smith's breasts—which the *Los Angeles Times* dubbed "freakish mammary glands"—were a primary subject of her 1992 interview with Regis Philbin and Kathie Lee Gifford. Newsman Larry King badgered Smith about her rumored enhancements on his news show. "Did you have breast augmentation?" he asked. "Up or down?" By the time Smith was famous and widowed from a wealthy geriatric oil tycoon, it was perhaps no surprise that she credited her breasts. "Everything I have is because of them," she said.

Not unlike Victoria's Secret models, Smith and Anderson titillated men and sold women a quixotic yardstick to measure themselves against. The popularity of these entertainers signaled to women that achieving a similar body type could bestow real power. Young women, already down on their figures thanks to the media and popular culture, wanted to fit this new definition of perfection. Many took to extreme dieting and surgery to achieve it. From 1990 to 1999, cosmetic plastic surgery procedures hit record numbers, topping one million each year. Anderson's body was the most desired by nonbombshells seeking surgical enhancement and by 1997 was "the number one requested body image in L.A.," according to a prominent plastic surgeon. Women begged to be cut up and reassembled to look like Pam.

Anna Nicole Smith, meanwhile, was the apotheosis of the image that marketers had so successfully peddled. Smith's rise suggested to many American women that becoming a sex object was their only path to power. But Smith's own pursuit of this path would come to a dark end.

Vickie Lynn Hogan of Mexia, Texas, wanted to be in *Playboy* so badly that she sent the magazine nude photographs of herself. Married with a son by the time she was eighteen, Hogan idolized Marilyn Monroe. After jobs at a fried chicken restaurant, Walmart, and a strip club, she was named Playmate of the Year in 1993. That was when America came to know her as Anna Nicole Smith.

Smith's voluptuous, unsculpted body was central to her appeal. Fans liked that she looked like she ate, especially in an age when beauty was often defined by emaciated models. Smith loved pizza and fried chicken. She confessed to never exercising and eating Godiva chocolates at every photo shoot. Her notoriety made curvy desirable again.

Smith resurrected an old body type and made it de rigueur: the buxom blonde. She supplanted the lithe Claudia Schiffer as the face of Guess Jeans, proving that excess flesh was the desired tool to market the country's most popular denim. Smith's curvy, five-foot-eleven, 155-pound, 36-DD frame was "uncontainable by ordinary clothes," opined the *Washington Post*. "She brings back visions of Hollywood glamour. We haven't seen that kind of charisma since Marilyn," said a Guess photographer.

But Smith's body would swell and then shrink over the years, making her a target. That she ate and didn't exercise first endeared Smith to America, but once she became overweight, she was mocked and shunned. The Marilyn comparisons ceased. Critics taunted her for not fulfilling the sexual fantasy that she had promised. She became a cautionary tale to young women that corpulence was undesirable, that they should nix the fried chicken if they wanted love.

Smith's pursuit of an unrealistic beauty ideal turned her into a carnival sideshow. "She spilled out of her tops, she spilled into the tabloids, she was a mess," the *Post* continued. She wasn't just mocked for her breasts; she was reduced to them. In *The Economist*: "There were only two of them, but they made a whole frontage: huge, compelling, pneumatic. They burst out of tight red dresses . . . or teased among feather boas, or flanked a dizzying cleavage that plunged to tantalizing depths. With them, a girl from nowhere . . . could do anything."

For Smith, marketing her sexually desirable body was an escape hatch from addiction, depression, and poverty. But eventually, the hatch slammed back on top of her. By the early 2000s, Smith had become so subsumed by prescription drugs, alcohol, and diet pills that a PR guru told *Adweek* she was a "walking catastrophe." Beginning in 2002, she starred in a reality television show about her varying attempts to maintain her body. *Slate* called *The Anna Nicole Show* a "druggy, delirious debut," and wondered if the channel was enabling an obviously intoxicated Smith, who could barely walk or speak coherently.

The deeper she plunged into addiction, the more she was mocked for being a pathetic floozy. Introducing Kanye West at the 2004 American Music Awards, a drugged-up Smith skims her silhouette with her hands, then asks the audience, "Do you like my body?" She seems vanquished after years of ricocheting between fame and antipathy. But she is still lucid enough to focus on the asset that had garnered her love and a career.

Critical response to the moment took two forms: it celebrated her newly slender physique obtained while consuming and endorsing a line of diet pills, or maligned her for being wasted. *Gawker* joked that a Mexican pharmacy association should commission Smith to open a drug store in her cleavage. *People* said she was "looking svelte." *Entertainment Tonight* reporter Kevin Frazier interviewed Smith before she took the stage. Afterward, he wondered on air if her behavior was a scheme to hawk her diet pills. "Something is going on. But at the same time, I wonder . . . is she crazy like a fox? Because is there any better self-promotion?"

Smith died three years later with eight prescription drugs and a heavy sedative in her system. After Smith's death, she was ceaselessly mocked for pursuing love and self-esteem from the outside in, even though it was exactly what society had instructed her to do. She saw no other option.

SEXY VILLAINESSES

There was only one type of woman routinely permitted to exercise power and show negative emotions in the 90s, and she wasn't even real. The sexy villainess was a regular on 90s television shows, adored for the havoc she wreaked and the cleavage she revealed. She undermined other women, poached their mates, and exacted revenge. Her real power only existed in her ability to conform to male desire.

Melrose Place perfected this villainess and her hobby, the catfight. What began as a "soft romance with girlfriends and boyfriends in the same courtyard" became an entirely new animal once Heather Locklear joined the cast, according to show writer James Kahn. Producers "thought they would shake things up and throw a bitch in the middle of this otherwise complacent thing we'd seen before," Kahn told me. Once the show swapped "bothersome Issues and Morals for infinitely more palatable Sex and Villains," as *Rolling Stone* put it, *Melrose* climbed to the top of Monday nights. Heather Locklear's cocky, sassy, self-made Amanda led the charge. Reviewers noted her "Barbie doll" figure and "scratch-your-eyes-out attitude."

The seductress was the only type of woman on the show. She came in three varieties: blonde, brunette, and ginger. Creators wouldn't allow anything to stand in the way of this model, even the law. Actress Hunter Tylo sued producer Aaron Spelling after he fired her from the *Melrose* cast because she was pregnant. Spelling's lawyers argued in court that Tylo's pregnancy would render her too fat to play a husband-stealing vixen, so she had to go. A jury awarded Tylo close to $5 million in damages. (Lisa Rinna, the actress who replaced her, announced her own pregnancy and continued working.)

Locklear's character, the temptress Amanda Woodward, ruled the show from the first episode she appeared in. She seemed to echo the kind of woman *90210*'s Brenda Walsh might become, but more boldly sexual, powerful, and blonde. Amanda was older, more promiscuous, and financially independent. She ran her own advertising agency and took any man she wanted, even if he was tied down. Amanda bought the apartment complex where the show's characters lived as a ploy to attract one of its residents. Then she demoted his ex-girlfriend, whom she happened to employ.

Stealing boyfriends and impeding female colleagues was only the beginning. "Evil Amanda" set temperatures boiling. One of the show's most infamous promotion posters simply read, "Mondays are a Bitch" over a closeup of Locklear's face. "She's not that bad. Her heart does beat on occasion, but then it just stops," Locklear once joked about Amanda. The *Los Angeles Times* classified her as a "stop-at-nothing female character who uses sex and smarts to neutralize men." Women viewers and other characters envied her. Men wanted to sleep with her and, at the same time, feared for their lives.

Journalists writing about Locklear contrasted how nice the actress was with the evil bitch she played on TV. Locklear knew how to schmooze reporters and score fawning news coverage. "The awful truth" about her is that "she is really very nice, the kind of woman who never speaks ill of her colleagues and refuses to trash her ex-husband," a *New York Times* profile explained. "It's fun to play things that you're not," the actress insisted. Still, Locklear tried to create distance from Amanda. "I love it. As long as they don't call *me* a bitch," she told *Entertainment Weekly*.

Melrose's Amanda and fiery Kimberly Shaw (played by Marcia Cross) embodied the trope of the empowered bitch, according to Kahn, who says that casting women as powerful, manipulative villains counterintuitively indicated progress. He likens it to old Hollywood when black actors were not given dynamic villain roles because of the concern of perceived racism: "You had to walk on eggshells. But finally, they were accepted and given really good bad-guy roles."

Kahn says women's evolution in Hollywood mirrored this progression. "Women were helpmates. They were sweet, and the love objects. You couldn't really make a woman a bad guy. But what it felt like was happening in the 90s, and what *Melrose* was certainly doing, was the same kind of parallel thing. We gave women power and showed that they could be bitches and that they can be horrible and manipulating, just like men. It was that, and in personal relationships, power was wielded to a certain extent with sex and emotional manipulation. That was a more classic way that women and their roles and domain was more emotional and psychological."

Creators of *Melrose* and other shows may have believed they empowered women by making them bitches. But surely, thwarting other women and stealing men wasn't new or empowering behavior for female characters on television. Soaps had long been beating that drum. *Melrose* employed an old daytime formula for primetime. All the women waxed villainous, crushed other women when they could, and fought over men.

For the writers, about half of whom were female, manipulator characters like Amanda and Kimberly were the favorites because their motivations and story lines were so rich, Kahn recalled. "Amanda was the bitch you wanted to be, the bitch positioned for greatness because she was freed of the convention of caring what other people thought. Her modus operandi and meaning of life was about winning at all costs," he said. Scripting women's histrionics was catnip for viewers, and was designed to spark watercooler talk the next day.

Another way to see it is that *Melrose* fed viewers a deceptive premise—that women's real power only existed in a world where sex was a weapon, and that their careers and achievements were merely props in vengeance plotlines or vehicles for enacting revenge. This myth was escapist at heart, like the sparkling shops and beach bodies of *90210*, but it was also a hollow promise to women. Both *Melrose Place* and *90210* seemed to claim power and space for new kinds of girls and women who were beautiful, mouthy, and free to speak their minds. But what these shows really did was propagate an old formula for women on television. Women could be cunning, feisty, and smart, but only when it came to scoring men or getting revenge on other women.

By 1997, women characters on television were more likely than men to obsess about their appearance, and to use precious dialogue (particularly in short sitcoms, where each line must reveal something about a character or move the action forward) to receive or make comments about their looks. They were also far more likely to be shown grooming, shopping, or maintaining their appearance. Women were rewarded for these strides, as shows like *Melrose* and *90210* had made clear. Women characters in media were more likely to discuss romance or their looks than be shown in jobs or at school, according to a report by advocacy group Children Now and the Kaiser Family Foundation. Amanda and her cohort on *Melrose Place* reinforced to women and girls that their power was limited to sex, violence, and consumerism.

BITCH SLAP

In May of 1992, MTV premiered a series that seeded television's most bitchifying and prolific genre, the kudzu of cable, which still reigns today: reality television. *Real World* is MTV's longest-running series to date, and it is still going strong. The formula—casting young, attractive roommates to argue, share toothpaste, and hopefully fuck, all on camera—forged the retrograde female stereotypes that still haunt reality television.

At its start, *Real World* was groundbreaking, casting diverse Americans and covering topics infrequently elevated elsewhere, like AIDS, abortion, and racism. But it wasn't long before the show's capacious treatment of humanity and social issues devolved into ugly clichés. In particular, *The Real World* created the villain bitch reality television couldn't live without. Women presented on *The Real World* were young, single, and educated. But their independence eventually became tethered to tropes like the bitch, slut, backstabber, and gold digger. These stereotypes were all discovered, cast, and nurtured on *The Real World* before they flourished on countless other programs.

Perhaps the most iconic female character of early episodes—and the one who was most often called a bitch—was Tami Roman (née Akbar) of *The Real World: Los Angeles*, which aired in 1993. During the show, she got an abortion, wired her jaw closed to drop pounds, and succeeded in booting off a fellow cast member, David, for sexual harassment. These experiences made her "the most screwed-up Real Worlder ever," according to one review. She and two other female cast members had approached producers to complain about David, claiming they didn't feel safe with him in the house. But reports blamed Tami for egging him on. Although Tami volunteered with AIDS patients and sparked a dialogue about abortion and body image, she is remembered most as a villainess for her perceived manipulations. In 1993, she told the *Los Angeles Times* that she was commonly stopped in shopping malls and asked, "Why did you kick David out of the house? . . . I mean, you acted like such a bitch, and now you seem so nice."

The bitch became one of *The Real World*'s best-recognized roles for women. Producers emphasized that they weren't looking to cast "stereotypes," but hopefuls auditioning knew better. "I could be the bitch," twenty-three-year-old Christine Sclafani told a *Miami Herald* reporter at a tryout in 1995. "I'd be the one arguing with everyone." Another viewer liked *The Real World* because it was "just like those live cop shows. Real people fighting all the time. One woman even admitted to being a bitch on the show."

When Melissa from the Miami season opened a cast member's mail, he called her a bitch, on camera. Another Miami character, Flora, called herself a bitch. A third female character that season, Cynthia, was "cute and sassy, but a two-faced backstabber," according to one paper. These women all appear in one season, suggesting that catfights were a narrative priority.

By 1998, a *Real World* "bitch" was so normalized that even being a victim of violence didn't win her sympathy. One of the show's most discussed episodes featured a male cast member slapping a female cast member in the face during the Seattle season. The so-called "bitch slap" was named among VH1's "40 Greatest Reality TV Moments." *Entertainment Weekly* ranked it atop a TV roundup because it was refreshing to see "what happens when people stop being polite and start being real."

During the 1990s, *The Real World* wrote the playbook designating the limited ways women could be portrayed on television, and also in real life. It planted and nurtured the villainess bitch in the documentary form the way *Melrose Place* had in fiction.

The show's very title promised these women were not just television characters, but real women commonly found in the world. "Reality TV isn't simply *reflecting* anachronistic social biases, it's resurrecting them," author Jennifer Pozner wrote in her critique of the genre, *Reality Bites Back*. "The genre has . . . created a universe in which women not only have no real choices, they don't even *want* any."

90s WITCHES

While 90s villainesses in nighttime soaps and *The Real World* commanded authority through sex, bitch slaps, and catfights, the supernatural realm afforded women characters actual power through magic. Paranormal women characters have long had more leeway than regular women to be messy, biting, and even unhinged. This tradition dates back to *The Wizard of Oz*, *Sleeping Beauty*, *I Dream of Jeannie*, and *Bewitched*. But most 90s witches and evil-thwarters wielded sex appeal alongside their spells, which they often cast to snare men or inflict revenge.

Sexy aliens and sorceresses skewed younger in the 90s to appeal to the sought-after teenage girl demographic. *Buffy the Vampire Slayer*, *Sabrina the Teenage Witch*, and *The Craft* took the discovery of female power to the consummate hell for outsiders: high school. Sarah Michelle Gellar's Buffy and Melissa Joan Hart's Sabrina were not only enchantresses experimenting with the contours of their power, they were also teenagers learning how to grow up.

Buffy, "the Chosen One," offs vampires, while Sabrina discovers that she is a witch on her sixteenth birthday and tries to use her powers for good. They also face mundane teen dilemmas—

like trying out for cheerleading and working a summer job at a coffeehouse—that can accentuate their otherworldly ones. Sabrina casts simple spells to change her outfit or put bullies in their place. In the 1996 cult film *The Craft*, witchery unites four outcast girls who use their powers to heal themselves and avenge wrongs perpetrated by jerky classmates. Both Buffy's and Sabrina's rich interior lives and motivations are complex and unfold over many seasons. *The Craft* confronted real problems like acceptance, abuse, and bullying. But viewers were constantly reminded, simply by looking at the witches, that their power and uniqueness came in a sexy, ebullient shell. Buffy and Sabrina are perky blondes, while *The Craft*'s characters wear midriff-baring shirts, short skirts, and thigh-highs.

When *90210* and *Melrose* godfather Aaron Spelling finally got a hold of a witch series in 1998, sex appeal eclipsed magic. *Charmed* follows a coven of witches, the Halliwell sisters, played by Shannen Doherty, Alyssa Milano, and Holly Marie Combs. Casting Doherty as a witch became a laugh line for many critics, implying that her public misbehavior and inherent bitchiness had won her the role. The Halliwell sisters fight evil in smoldering eye makeup and plunging tops too risqué for Sabrina or Buffy. They are described as "beautiful," "fashionable," and "feminine" witches. They may rid the world of warlocks, but the apogee of their power is their appearance. Interspersed with their spells—which mostly occur in their home, relegating them to domestic space—is their romancing "a procession of attractive males," wrote the *Los Angeles Times*. "*Charmed* is a perfect postfeminist girl-power show," Milano said in an interview. "These women are strong, but they're still feminine and accessible." Episode titles include "Dead Man Dating" and "Love

Hurts." Buffy, too, hunts for love, and "takes out the undead between dates," according to the *Chicago Tribune*.

Nineties witches often used their powers to either find love or exact revenge, much like the nonmagical Amanda from *Melrose Place*. In the 1998 film *Practical Magic*, sisters Sally and Gillian (Sandra Bullock and Nicole Kidman, respectively) tamp down their sorcery until they need it "to overcome the obstacles in discovering true love," according to the film's marketing copy. Witchcraft is therefore only "practical," as the title suggests, when it's used to catch a mate.

In *The Craft*, Sarah (Robin Tunney) gets revenge on a classmate who humiliates her by casting a spell that makes him fall in love with her. Nancy (Fairuza Balk) later tricks him into sleeping with her by making him think that she is Sarah. Nancy then kills him for rejecting her when he learns the truth.

Too much power drives the witches crazy and forces them to turn on each other in a series of catfights. Sarah practically battles Nancy to the death after Nancy abuses her power by killing people. The film suggests that women's extreme power should be constrained or else it will be abused, and that women should be responsible for policing one another. Buffy's abilities, too, need controlling. Her initial access to her power is shepherded by a male authority figure, the fuddy-duddy school librarian, Giles, who is dubbed "the Watcher." Buffy's otherworldly abilities are not only marshaled by male authority, they are also chastened by it. She is expelled from high school for causing trouble when she was really just trying to save the world. On her way to destroy evil demons, she is stopped by the school principal, who thinks she's cutting class.

When women in supernatural settings rejected catfighting

with other women or exploiting themselves sexually, they were seen as nags. On *The X-Files*, then television's longest running sci-fi series, FBI agent Dana Scully (Gillian Anderson) had no magical powers. But the show's plotline, which involves a series of alien and paranormal investigations, allowed for Scully to be a complex and spirited character more dynamic than most women on television. And yet, she is the resident skeptic, charged with reining in her partner's at-times irrational belief in extraterrestrial life.

Scully is the dogmatic realist, demanding that investigations be fact- and evidence-based. Her partner, FBI agent Fox Mulder (David Duchovny), dreams of aliens and throws caution to the wind. Ultimately, the show reveals that he is right, aliens do exist, leading Scully's tightfisted caution to seem annoying and overblown. Anderson, who portrayed Scully starting in 1993 and for all of the show's ten seasons, said the network had wanted a sexpot Scully—"taller, leggier, blonder, and breasted"—for the role. Maybe if Scully had been hotter, she wouldn't have been such a harpy.

2

SEX IN THE 90s

Many look back at the 90s and remember sex. What might be less memorable is how the sex differed depending on your gender. Sex in the 90s was characterized by women being blamed and shamed, and by men being celebrated. This began as early as grade school. Abstinence-only sex education was the predominant school-based curriculum.

In 1996, abstinence-only became the national standard when it received a multimillion-dollar windfall. This was done, oddly enough, through welfare reform. The legislation that President Bill Clinton signed into law that year incentivized abstinence-only programs in America with an appropriation of $50 million per year. Abstinence-only programs taught children and teens that having sex before marriage was socially unacceptable and that avoiding sex was the only real way to prevent pregnancy and STDs. The law's definition of abstinence-only education warned that sex before marriage caused "harmful psychological and physical effects." Lessons included how to "reject sexual advances" and taught that drugs and alcohol "increase vulnerability to sexual advances." In other words, the lessons stopped well short of practical sex education, such as how to use birth control and how to have a healthy sexual relationship.

My own sex education experience in middle Georgia in the 90s hewed to the abstinence standard, and dispensed fear of HIV/AIDS and STDs as a substitute for birth control. My middle school class filed into the auditorium to hear a lecture from a prominent HIV/AIDS expert in the Southeast who not only offered grave warnings, but also provided supposed photographic proof of his patients' suffering. He projected their open sores, bloody limbs, and rotting sex organs onto our two-story-high auditorium wall. We exited the assembly stiff with fear and ready to skip lunch. It was a common tactic in abstinence-oriented sex education. "Teachers say, 'Let's gross these kids out so much they won't want to have sex,'" a seventeen-year-old student named Annie told a *Washington Post* reporter in 1999. "But then we don't remember the information." Anything unrelated to disease catching or baby making was unspeakable.

As it turns out, abstinence-only sex education is ineffective, not just in America, but all across the globe. It doesn't prevent sex, and it surely doesn't stop the spread of disease, according to legions of policy and healthcare experts and studies. In fact, abstinence education is often blamed for the high rates of teen pregnancy and sexually transmitted diseases that struck the US in the 90s and early 2000s. And so, sex education in the 1990s wasn't about sex. It was about control. Abstinence education forced girls to go on the defensive, guarding against sexual overtures from boys, while absorbing blame for any consequences. It also neglected non-heterosexual experiences.

Disease as punishment for unsanctioned sex was a common theme in abstinence-oriented sex education. Becoming a sexually active woman in the 90s was colored by the fear of contracting HIV/AIDS. The need for fear was real. By 1994, the disease had become the number one killer of twenty-five- to forty-four-year-olds in the United States. More than 440,000 cases had been reported, and more than a quarter million people had died of the disease.

Men were in the majority. But women were undercounted due to lack of participation in clinical trials and a narrower clinical definition of the disease before 1993. In 1996, women comprised 20 percent of all cases. Many girls' fears were stoked by the 1995 film *Kids,* in which a teenager played by Chloë Sevigny contracts HIV after having sex just once. On television, teen sex that *didn't* trigger repercussions (particularly for women) was widely frowned upon. When Brenda lost her virginity to Dylan on *90210,* "the affiliates were scandalized—not because they had sex, but because Brenda was happy about it, and it didn't have any dire consequences," creator Darren Star told a reporter.

As the decade progressed, fear of sexually transmitted infections other than HIV/AIDS crescendoed because more people were getting them. More than three million teenagers were contracting STDs each year, according to a 1995 report by the National Commission on Adolescent Sexual Health. It noted that European teenagers experienced much lower rates of STDs and pregnancy because "European countries tend to be more open about sexuality, and their official governmental policies focus on reducing unprotected intercourse, rather than reducing sexual behaviors." The report recommended scrapping abstinence-only education, but that didn't happen. By the end of the decade, fifteen million Americans were contracting STDs each year, and two-thirds of the infected were under twenty-five.

Women became de facto custodians of sexual health and safety during this outbreak. They were blamed and shamed for contracting infections like chlamydia, gonorrhea, or syphilis. This was in part because their symptoms and infections were less obvious than men's. It was suggested that they could trick unsuspecting partners. Women also had more to lose, since untreated infections could cause infertility or cancer.

Reports about the rise of STDs in the 90s often featured more young women being interviewed about the diseases than men, which not so subtly reinforced the idea that women were at fault for transmitting them. One story in the *New Orleans Times-Picayune* attributed the spike in STD rates among teenage girls in the area to the increasing number of them sleeping with older men called "Cat Daddies." These men extend "promises of jewelry, snappy shoes and fancy nails and hair in return for sex," the paper wrote. An expert at the Centers for Disease Control and Prevention said that young girls would need "special attention" when it came to STDs, specifically because they dated older men like "Cat Daddies."

Meanwhile, it was collegiate women who hustled to campus health centers to get tested, not their male counterparts. Women saw iconic female television and film characters like Kim Cattrall's Samantha in *Sex and the City*, Tiffani Thiessen's Valerie in *90120*, and Janeane Garofalo's Vickie in *Reality Bites* seek out HIV testing, but only after they exhibited sexual promiscuity.

Seeing these beloved women test themselves marked new moments of cultural awareness about STDs. But it also signaled that wantonness had finally caught up with these women and that their freewheeling sexuality had a cost. Why didn't we see their male partners getting tested, too? And why weren't other STDs as discussed as HIV/AIDS? "There are no Magic Johnsons for gonorrhea," the *Washington Post* glibly observed.

SEX IS HOT HOT HOT

In 1999, STDs reached epidemic levels in America. MTV teamed up with the Kaiser Family Foundation to find out who got them and why. The channel aired the results of the study as part of their *Sex in the 90s* series. But the program seemed to do more to stoke fear than to educate. Likened to "a detention-class film strip" by *Entertainment Weekly*, the special shamed those who contracted STDs, especially the women.

"Sex is famously hot hot hot, but nobody talks a lot about how badly you can get burned," warned MTV News host and father figure Kurt Loder. "Misconceptions" and "careless behavior" caused diseases that "most people know little or nothing about," he intoned. "As clear as this danger is, many people continue to make tragic mistakes that can't be undone."

The film doesn't hold politicians or educators accountable, or demand more resources for sex education. Instead, it heaps blame on the regretful, infected interviewees—almost all of whom were women. The three straight women interviewed told a familiar story. After unprotected sex with a trusted partner, they were now "paying the price" with chlamydia, herpes, gonorrhea, etc. They accept responsibility for "making a mistake," and admit they deserved the "slap in the face" diagnoses.

Women and girls had more to worry about than physical discomfort. An STD diagnosis meant not only that they could become sick, but also that they would inevitably be labeled sluts. "I had that connection between promiscuity and STD. Once I got an STD, therefore, I must be promiscuous," said Adina in the MTV film. A fire roars behind her, a reminder of how she "got burned."

"'I have to be one of the biggest sluts in the world' is what was going through my head because this would not happen to somebody who doesn't sleep around," added another woman, Kari. A third woman scolded viewers, "'It's not going to happen to me'—that's the biggest myth. It will happen to you."

Jennie Miller was ashamed to discover she had human papilloma virus (HPV) and told ABC News, "The first thing that came into my mind was, 'I am never going to have sex again . . . I'll never date again.'" Women were expected to be sexual gatekeepers, required to set boundaries for going to bed, and blamed if things went awry. But even when women took the precautions to prevent STDs, they were often slut-shamed. An executive producer of *Beverly Hills, 90210*, Jessica Klein, said she was stopped by higher-ups from doing a scene in which a female character puts condoms in her purse before a date. "They said that would make her seem like a 'slut,'" Klein said.

TRAUMARAMA

In the epoch just before the mainstreaming of the internet, teen magazines offered the biggest window into ideas and conversations about how girls felt about sex and their bodies. *Seventeen* brimmed with front-of-the-book features dedicated to answering readers' many queries about these topics. Today, if you're a girl with a question, you just Google it. Back then, as strange as it sounds, you wrote in to a magazine for advice, or hoped a magazine advice columnist would answer a question that was brewing in your mind. The gurus at *Seventeen* knew the answers and delivered them with expert confidence, if not actual expertise.

Readers posed queries about tough topics like STDs and virginity, and often it was the magazine's staffers, not medical doctors or psychologists, who would reply. Sometimes their "advice" was pejorative and judgmental. In the January 1995 issue, a reader inquired about crabs. Before explaining the parasitic infection, the magazine replied that crabs sounded made up, "but they're real. And they're real gross." Another woman writing to *Seventeen* in May that year asked if her inverted nipples were normal, to which the "expert" responded, "You just have to deal with those first couple of times being shirtless around someone else and eventually your embarrassment will go away."

Features like *YM*'s "Say Anything" and *Seventeen*'s "Traumarama" published real-life embarrassments, such as period-related humiliations, couples kissing and locking their braces together, and teachers reading confiscated love notes, stoking fear among teen girls that just *being* a girl guaranteed mortification. They reinforced the "grossness" of girls' bodies, period-shaming, and the injunction that girls should please boys. Quizzes like "Are You a Boyfriend Addict?" "Are You Obsessed with Him?" and "Are You Jealous?" extended the theme.

These magazines also offered some fascinating reporting on topics like censorship at student newspapers, the real lives of teenage moms, and useful service journalism, like what kind of summer job might suit you best or how to boost your SAT vocabulary. But this fare was less common. I recall as a teenager skipping right past that stuff as I searched for advice that would help me better navigate being a teen girl. Instead, what I read only fed my insecurities.

Teen magazines may have shamed young women's bodies and sexual curiosities, but they spelled huge opportunity for advertisers. New reads like *Teen People, Jump, Twist,* and *All About You* raced to the shelves to capture a sliver of the burgeoning market. By 1998, *Seventeen, Teen,* and *YM* circulation numbers hovered around two million per magazine. In the first quarter of that year, *Seventeen* earned $17.5 million in ad revenue. *YM* revenue topped $7 million, and *Teen* brought in over $6 million. *Teen People* quickly overtook staples *Vogue* and *Vanity Fair* in circulation upon its debut.

OBJECTIFICATION THEORY

In 1997, researchers Barbara Fredrickson and Tomi-Ann Roberts offered a theory about the underlying cause of girls' insecurities, lack of self-esteem, and body strife. "Objectification theory" explained that women and girls are "acculturated to internalize an observer's perspective as a primary view of their physical selves." Thus, because society values female bodies primarily for their function and consumption, women and girls are more susceptible to suffering as their bodies change, like during puberty, but also due to pregnancy, weight fluctuation, and aging. This objectification enables discrimination, sexual violence, undervaluing women, and depression, the authors wrote.

In other words, this hall of fun-house mirrors in which society objectifies girls, then hoodwinks them into objectifying themselves, causes women and girls to experience real-world suffering.

In the absence of an authentic playbook for female identity and sexuality in the 90s, girls coming of age understandably turned to media and cultural scripts that prized sex and objectification. What girls were told about sex—in school and society, and by elders and peers—stood in stark contrast to what girls were *sold* about sex through entertainment, advertising, and pop culture.

This knowledge void, combined with the struggles of self-esteem and body image created or exacerbated by the media, haunted girls and young women. Advertising and entertainment increasingly sold young women the dictum that their sexuality should be bionic, breasty, and blonde, and that this was achievable through consumption: buy this diet or this underwear to make men want to sleep with you, because being desired by men is the path to self-esteem, power, and love. These pronouncements underpinned advertisements for practically everything in teen magazines and on TV and movie screens. Consumer-oriented messaging about sex would evolve into raunch culture of the early 2000s (see Chapter 12), which inveigled women into believing that feminism and independence were won only through overt, male-engineered sexuality.

Thanks in large part to marketers and educators, teenage girls and boys were imprinted with radically different attitudes toward sex. Boys were encouraged and even pressured to pursue sex (with girls and girls only). Girls like me learned about sex through a scrim of fear, but we were also charged with preserving the sanctity of our own bodies, which we didn't yet know or understand. Boys' sexual aggression was celebrated; girls were taught submissive sexuality, and blamed and shamed for sexual consequences.

COERCED AND FORCED SEX

In 1994, sociology professors John Gagnon and Edward Laumann published a study of sex in America that was widely called the most definitive to date. Two hundred twenty researchers conducted in-person interviews with nearly three thousand five hundred men and women between the ages of eighteen and fifty-nine. Perhaps the most worrisome finding was that more than a third of young women reported that peer pressure convinced them to have sex for the first time. The study authors suggested that young women should learn to resist peer pressure as one part of avoiding teen pregnancy. It was as if individual women needed to answer for and fix a broader cultural problem.

To wit, girls' early sexual experiences were often blurred by substances not uncommonly given to them by boys in the first place. "Whatever gauzy, Fabio-induced fantasies they had about the first time were usually numbed out with drugs or alcohol; the reality was, at best, disappointing, at worst, coerced or forced," reported Peggy Orenstein in her 2000 book, *Flux: Women on Sex, Work, Love, Kids, and Life in a Half-Changed World.*

But the conversation about sexual consent was shaped by the concept of date rape. "Date rape" captured the imagination in the early 90s, and led to a backlash against women who discussed it or, worse, accused men of it. This was thanks largely to the William Kennedy Smith and Mike Tyson rape trials, and also to journalist Katie Roiphe's book *The Morning After*, which questioned whether date rape was as prevalent on college campuses as it seemed from feminist conversation. Reports about rape's pervasiveness worked to debunk the myth that women are mostly raped by strangers; the crime is perpetrated more often by men who know their victims. However, knowing the perpetrator led some women to blame themselves or to question whether what they experienced was technically rape. This reaction dovetailed with the chauvinistic theory that many women who charged rape were making it up. "Men say it is a concept invented by women who like to tease but not take the consequences," reported *Time* in 1991.

Still, to stanch the flood of sexual assault, feminists, campus activists, and victims' advocates marched, protested, and organized Take Back the Night walks. Detractors accused them of being hysterical and classifying too much sex as rape. When you consider that one in four women are reportedly raped in their lifetimes, maybe they weren't hysterical enough.

This boorish attitude prevailed in the sexual assault scandal and cover-up that embroiled the military in 1991. Eighty-three women and seven men were sexually assaulted by navy and other military personnel during a convention of the aircraft carrier support group, Tailhook. Navy lieutenant Paula Coughlin told *60 Minutes* that she was sexually assaulted by a horde of drunk men at the Las Vegas gathering. She said she feared being raped, and was subsequently ignored when she reported the incident to her superiors. A Pentagon report later found that the navy had thwarted its own investigations into the assaults at Tailhook because of inherent organizational misogyny and a desire to save face. It cited an internal dispute during which one admiral compared women navy pilots to "go-go dancers, topless dancers or hookers." Another high-ranking official added that the victims "would not have gone down the hall if they did not like it." The Tailhook incident furthered the belief that no woman was safe from victimization or blame.

THE CONDOM QUEEN

Against this backdrop of abstinence, disease, and sexual coercion and violence, Surgeon General Joycelyn Elders embraced talking about sex with gusto. From the outset of her tenure, the first-ever African American and second-ever woman surgeon general advocated for widespread sex education in schools. She also championed the availability of contraceptives. Elders talked frankly about safe sex to prevent unwanted pregnancy and disease. Instead of flowers, she kept a bouquet of roses fashioned from condoms on her desk. Being disparaged as "the condom queen" didn't bother her. "If I could get every young person who is engaged in sex to use a condom . . . I would wear a crown on my head with a condom on it!" she said.

In December 1994, at a United Nations conference on World AIDS Day, a psychologist asked Elders whether educating students about masturbation could limit the spread of HIV/AIDS. She answered that sex education should begin "at a very early age." She continued that masturbation "is something that's part of human sexuality and it's part of something that perhaps should be taught. But we've not even taught our children the very basics. And I feel that we have tried ignorance for a very long time and it's time we try education."

This reasoned answer—that there's nothing wrong with

self-pleasure, and that comprehensive sex education must be the nation's priority—was not the takeaway. Critics accused Elders of wanting to teach schoolchildren *how* to masturbate. Even though Elders tried to correct the record and clarify that she hadn't been recommending that schoolchildren take masturbation tutorials between geography and lunch, only explaining that masturbation was a natural part of life and as such might be included in comprehensive sex education, she was pilloried by the media. Late-night television mocked her for talking about sex. Jay Leno joked that she "pardoned Pee-Wee Herman," in reference to actor Paul Reubens, who had been arrested for masturbating in an adult movie theater. Conan O'Brien cracked that she would return to teaching classes "filled with teenage boys." David Letterman mocked her on numerous episodes. In a Top Ten list of least-convincing alibis, he said number two was being "alone in my room doing some of that Joycelyn Elders stuff." Late-night comedians determined that Elders was too sexual for men but plenty qualified to teach boys. As ever, their jokes belied the double standard that sexuality was forbidden for women to discuss but was men's province to flaunt.

Elders was right, of course. The country's abstinence-oriented sex education was failing to prevent disease or cultivate sexually healthy adolescents. Thoughtful, comprehensive sex education could help. But Elders's comments exposed the shame inherent in sex, shame that was baked into abstinence-only education's legal definition: "Sexual activity outside of the context of marriage is likely to have harmful psychological and physical effects."

According to her biography on the National Institutes of Health website, Elders simply "left office" in 1994. There is no mention of why or how. In fact, her dismissal was almost gleeful. The White House "went out of its way to make clear that her resignation had not been voluntary," but, rather, was "forced," according to the *New York Times*. "If she had not resigned, she would have been terminated," said Leon Panetta, the White House chief of staff. The message was that she was bullish and rogue and could not be controlled. "Getting Out the Wrecking Ball" was the title of a *Time* magazine article about the firing.

Elders's male predecessor, Surgeon General C. Everett Koop, who served Republican presidents Reagan and Bush, had discussed masturbation, too, and in the exact same context as Elders—AIDS prevention. He somehow managed to keep his job. "I went to Washington feeling like prime steak, and I left feeling like low-grade hamburger," Elders told a reporter.

HORN DOG

Heterosexual male sexuality was cause for celebration in the 90s. While bodily shame and blame for disease stuck to women, men faced no such consequences. Instead, their libidos were enshrined in the media, laws, medicine, and even the White House.

Men's sex lives were extended for decades by one particular 90s discovery. In March 1998, the FDA approved an erectile dysfunction drug, sildenafil. Soon after, a small blue pill called Viagra crashed date nights and relationships across the country. The drug was an instant hit—some forty thousand prescriptions were filled during its first weeks on the market. Viagra was met with plenty of jeers and jokes about aging male virility. However, it also came to symbolize the superiority of male sexuality over female sexuality. Where was the drug that would help women achieve pleasure in the bedroom? It still doesn't exist. Influencers believed Viagra was the answer to women reaching for power in the 90s, and predicted the drug would stop them cold. Bob Guccione, the publisher of *Penthouse* magazine, assured men the drug would "free the American male libido from the emasculating doings of feminists." In short, sex was power. Now, men could get it in pill form.

Viagra debuted shortly after America learned that the president of the United States had an affair with an intern. The celebration of male sexuality in the 90s is inseparable from Bill Clinton. By the time he became president, there was both a familiarity and a fatigue with his caddishness, said Ty West, who covered the scandal for the NBC News magazine *Dateline*. "Being a horn dog was part of his persona before he was elected. So it was like, we've already voted on that. This is not new," West told me. Monica Lewinsky herself had mocked Clinton's seamy reputation with women before she became involved with him, parrying jokes about a colleague needing "kneepads," and disagreeing with female White House staffers who found their boss irresistible.

If male sexuality is prioritized, presidential sexuality is cause for genuflection. The media and some of Clinton's own staff excused his "horn dog persona." That Clinton was just being himself by having an affair with a White House intern helped shift the blame for the sex scandal onto Lewinsky. The president's public-facing discretion on the matter was rewarded by the culture. This took many forms. There were bro-y maxims printed on buttons and bumper stickers like "My President Slept with Your Honor Student," "Commander in Heat," and "One More Whore and We Get Gore." People wore shirts with the president's photo that read, "Will Work for Head." A *New Yorker* cover featured a slew of news microphones trained on Clinton's crotch, reminding America how newsworthy his sexual magnetism had become.

It even spilled into his dealings with the press. Nina Burleigh, a White House reporter, wrote about the thrill of Clinton ogling her during an Air Force One card game. "It was riveting to know that the President had appreciated my legs," she wrote. "If he had asked me to continue the game of hearts back in his room . . . I would have been happy to go there and see what happened . . . I probably wore the mesmerized look I have seen again and again in women after they have met him."

Reporter Tabitha Soren interviewed Clinton a handful of times in the six years she covered him for MTV News, and recalled his comments about her appearance during interviews. "He'd say, 'That's a great skirt.' Or, 'I really like that color on you,' something slightly innocuous, but that makes you aware of your body and you get off track. It's unsettling, even if it's a compliment." Soren believes the fawning was tactical, intended to distract her rather than flirt. "After it happened a couple of times I realized, he's doing this on purpose. I'd walk out and hit my forehead like Homer Simpson. He did it again! How could I waste a minute answering him?" she told me.

ORAL IN THE OVAL

The newsworthiness of Clinton's sex drive quickly became a referendum on oral sex. The 90s sex education vacuum fed curiosity and misconceptions about fellatio, so a public debate on the sex act seemed most welcome. The theory that emerged was that men deserved and should be rewarded with oral sex, and women should be pleased to oblige.

Independent counsel Kenneth Starr's 1998 report of Clinton and Lewinsky's affair reads less like a political or legal document than it does like a bodice ripper. Starr cites instances of "oral-anal contact," and attempts to poeticize a scene of Lewinsky fellating the president by describing a ray of sun landing on her face.

On the heels of the maniacally detailed Starr report, Lewinsky's name practically became shorthand for "blow job." In the 2002 HBO documentary *Monica in Black and White*, a man asks her, "How does it feel to be America's premier blow job queen?"

"I don't actually know why this whole story became about oral sex," she answers. "It was a mutual relationship."

One theory is that America hadn't collectively confronted heterosexual oral sex until it became presidential. Oral sex, particularly the kind received by men, became a barometer for sexual proclivities and appetites. Even though it was consecrated in pornography, whether or not the act qualified as sex became hotly debated. The president didn't think it was, and he was the leader of the free world, so maybe it wasn't.

Thanks in part to abstinence-only sex education, teens and preteens experimenting with oral sex also believed that it wasn't actually sex. In 2000, the *New York Times* published a hand-wringing report blaming the uptick in sexual activity, like oral sex, among younger kids on divorce rates and absent parents. One interviewee added rap music and MTV to the list of culprits. A psychologist interviewed said her preteen clients called themselves virgins saving themselves for marriage, but "they've had oral sex 50 or 60 times. It's like a goodnight kiss to them, how they say goodbye after a date." Abstinence-only education had convinced kids that oral sex was a work-around—they could avoid unwanted pregnancy and disease (though the latter wasn't true) and still have fun. Among middle and high school cohorts, including my own, it was a way to skirt virginity loss, or a carrot for someone you weren't ready to sleep with.

The 1994 study about sex in America found that only three sexual practices were "appealing to more than a small fraction of people": vaginal intercourse, watching a partner undress, and oral sex. They found the last was most likely to be enjoyed by men, not women. While few women over fifty had performed oral sex, three-quarters of women under thirty-five had.

Teen media reinforced male primacy in sex and female victimization. Researcher Laura Carpenter looked at two decades' worth of *Seventeen* magazines to determine which sexual scripts it normalized for teen girls. She published her findings in 1998, the year of the Clinton sex scandal, and found that the magazine's first mentions of oral sex were two articles disparaging fellatio in 1994. Cunnilingus wasn't mentioned at all, obliterating its possibility. Magazine editors "presented boys who wanted casual sex as dangerous but deviant and depicted girls who were abandoned after sex as unlucky individuals who chose the 'wrong' boys."

This idea of the primacy of male sexuality expanded to kids and teens who, in the 90s and early 2000s, were learning about sex through unprecedented access to pornography online. Pornography has promoted a restricted view of sex that persists today, said Cindy Gallop, a sex educator and the founder of the real-sex website Make Love Not Porn. "Because the vast majority of mainstream porn is made by men for men, the entire raison d'être of every single porn scene is to get the man off," she told me. "As a result, we now have an entire generation of guys and girls growing up learning that the entire raison d'être of sex is to get the man off."

Certainly, pornography's blow job fixation led scores of young men to believe they were entitled to the act, that they needn't reciprocate, and that all women loved to perform it. These misconceptions only spread as pornography exploded with the growth of the internet. And mainstream pornography, then and now, still prioritizes male orgasm and acts that achieve it. So it was no wonder that by 1998, when America learned of an affair in the Oval Office, the takeaway became the performance of blow jobs, derision for the giver, and backslaps for the recipient. It was the quintessential act that extolled male sexuality.

ENTITLEMENT AND FEAR

Reports of date rape and coercive sex, body shame, and blame for disease complicated sex for girls and women in the 90s and forced them to play defense and internalize consequence. This conflicted with what boys and men absorbed. Sex for them was there for the taking. Entitlement to sex influenced media depictions of famous men and sex acts and underpinned the spread of pornography online. It was theirs forever now that Viagra was in the picture. There was no better symbol for this than the gray-haired occupant of the Oval Office, whose sexual proclivities became legendary. Faulting girls for sex while lauding boys and men shaped how a generation came of age and discovered relationships and pleasure.

Thus, old, stale messages that women should fear the sex that men were entitled to persisted in the 90s, despite the decade's lip service to modern ideals about free sexuality and gender equality. The binary of entitlement and fear also filtered into media portrayals of each gender. It saturated how the press covered famous and infamous women, particularly when they were associated with sex or victimized. This dichotomy would shape the decade's two biggest sex scandals.

3

THE GOLDILOCKS CONUNDRUM

I t was in this climate of abysmal sex education, coercion, violence, shame, and blame that sex scandals replaced baseball as the national pastime. And the two scandals that bookended the decade—starring Anita Hill and Monica Lewinsky—illustrate how 90s culture forced women into a Goldilocks conundrum of female sexuality: there was only too hot or too cold, and no such thing as just right.

Ultimately, it wasn't what Monica Lewinsky did in the White House's inner sanctum that led to her bitchification. It was how she behaved afterward, and how that was translated by the press. She was unapologetic, sexual, and playful. She kept a semen-stained dress, sent love letters, filled her apartment with red roses, and giggled with Barbara Walters. She was blamed for not guarding herself against sex and for liking it. She was hated for not being attractive or thin enough to nearly torpedo a presidency, and for striving and taking too much. She was too hot.

In contrast, because she reported sexual harassment, Anita Hill was portrayed as stiff, frigid, and sexually unavailable. Her critics called her a scorned woman out for revenge. *Saturday Night Live* depicted sexual harassment as failed flirting and blamed the accused Clarence Thomas for failing to bed the frosty, snobbish employee. Hill was too cold.

Both were classic victims, blamed for powerful men's bad (and unlawful) behavior. The bitchification narrative muddied our view of them, however, preventing us from seeing them this way.

"A LITTLE BIT NUTTY AND A LITTLE BIT SLUTTY"

From today's vantage point, what happened to Anita Hill reads like a gallows humor satire of 90s bitchification. But there's nothing fictional about it. The plot is straightforward, if not entirely familiar to most: A former civil servant, who in the 1980s worked at the government agency charged with prosecuting sexual harassment cases, says that she was sexually harassed by her boss. A decade later, that boss is nominated to serve on the Supreme Court. The former civil servant testifies before the United States Senate about her former boss sexually harassing her, arguing that this makes him unfit for a job on the nation's highest court. But the senators humiliate and mock her. The media echo chamber amplifies the senatorial scolding. She is castigated as "a little bit nutty and a little bit slutty." Her life is more than a little bit destroyed, as a result.

Anita F. Hill grew up the youngest of thirteen children on a farm in Lone Tree, Oklahoma. Her chores included harvesting peanuts and digging crawdads from mud bogs. Hill first met Clarence Thomas, a Pin Point, Georgia, native who worked in government, at a Washington party in 1981. A mutual friend introduced them, both Yale Law School graduates, and they connected over a shared commitment to improving civil rights. "He struck me as sincere, if a little brusque and unpolished," Hill said in her memoir, *Speaking Truth to Power.*

Not long after, forty-three-year-old Thomas became assistant secretary for civil rights in the Department of Education. He offered the twenty-four-year-old Hill a job as his personal assistant. Though their politics differed—Thomas was a conservative, Hill a liberal—she had been looking to leave the big law firm where she worked and transition to a career in public service. "Perhaps this would be the dream job I had hoped for," she recalled thinking.

Coming from the private sector, Hill faced a steep learning curve. The familiar feeling of not belonging—a theme she'd come to know throughout her early life—reemerged. "Because I was a young, single black woman, the rumor mill speculated that I had been hired for both my race and my sex," she wrote.

Despite Hill's objections, her duties as Thomas's assistant included being an ear for his personal problems. He began asking her out, and pressuring her when she refused. Hill called Thomas "arrogant" and said he thought "his position entitled him to personal as well as professional access to his staff." She recalled that the more she rebuffed him, the more "vulgar" Thomas became.

Soon, he was spouting off NSFW sex talk and recounting acts he'd watched in pornographic films, like orgies and women having sex with animals. Thomas told her about a film star called Long Dong Silver and asked her, "Who placed a pubic hair on my Coke?" Reflecting on it in her memoir, Hill was certain that this behavior was a tactic to minimize her: "I was extremely uncomfortable talking about sex with him at all, particularly in such a graphic way, and I told him repeatedly that I didn't want to talk about these kinds of things. I would also try to change the subject. I sensed that my discomfort with his discussions only urged him on, as though my reaction of feeling ill at ease and vulnerable was exactly what he wanted."

A BRIEF HISTORY OF SEXUAL HARASSMENT

Sexual harassment has a long and tangled history in the United States. What's always been true is that women are overwhelmingly the victims, male bosses are chiefly the perpetrators, and sexual harassment is meant to disempower women in the workplace. This tactic dates back to slavery and was also rampant in the early American factories where young women worked. Victims of sexual harassment and rape were often punished or outcast, feeding the belief that women were to blame for tarnishing the male realm of the workplace with their sexuality.

"It was immoral for young girls to be working alongside men and subjecting themselves to the 'natural licentiousness' of the workplace," explained Alice Kessler-Harris, a women's labor historian, paraphrasing the widely held social view of the 1800s. A 1914 report found that a glass factory charged with widespread sexual harassment attempted to solve the problem by replacing all the young, sexually attractive women with older ones.

Like countless women before her, Hill wanted only for the behavior to stop. It was interfering with her ability to do her job, which she now regretted taking. "My stomach began to tie in knots at the thought of going to work each day," she said. Distraught, Hill confided in a friend, who suggested she change her perfume, as if that would somehow dissuade Thomas. The friend had meant to help, but her advice suggests how rudimentary the understanding of the problem was at the time. Back then, there was no tool kit. No one recommended Hill file a formal complaint. Hill was like many Washington newbies: the most powerful person she knew was her boss.

Sexual harassment didn't have a name, in fact, until 1974, when Cornell lecturer Lin Farley coined it. She used the term in a consciousness-raising group organized to protest the sexual harassment of Cornell University lab employee Carmita Wood, who quit her job to flee her tormentor. Farley testified before the New York City Human Rights Commission, using the term "sexual harassment" for the first time. Still, there was no legal recourse for victims. While in law school, the feminist lawyer and thinker Catharine MacKinnon devised a legal framework for prosecuting sexual harassment claims under Title VII of the Civil Rights Act. The Equal Employment Opportunity Commission (EEOC) acknowledged the problem of sexual harassment in 1980. MacKinnon's template was used and later codified in June 1986, when the Supreme Court heard its first sexual harassment case. *Meritor Savings Bank v. Vinson* was a particularly egregious example—a boss forced his employee to have sex with him repeatedly, insisting it was required of her—and the court ruled that sexual harassment, or discrimination on the basis of sex, was indeed a civil rights violation. MacKinnon warned that the court's decision wouldn't end sexual harassment, which was less about sex than an entrenched and successful means of workplace control. "It doesn't mean that women are always believed when they say they were harassed or that harassment won't happen anymore. But it does as much as the law can do," she said.

There was backlash, of course, even from feminists. When Helen Gurley Brown, the *Cosmopolitan* editor and indefatigable defender of carefree sex, was asked if any of her staffers had been sexually harassed at work, she answered, "I certainly hope so." She also recalled games of "scuttle" at the radio station where she had worked, in which men chased attractive women around the office until they caught them and tore off their panties. In preparation, girls wore their best panties to work.

COMING FORWARD

The EEOC is the federal agency that collects and investigates workplace sexual harassment claims. It defines sexual harassment as "unwelcome sexual advances, requests for sexual favors and other verbal or physical harassment of a sexual nature." In 1982, Ronald Reagan tapped Thomas to run the EEOC. The irony was not lost on Hill. However, she chose to continue working for Thomas, a move she describes as a matter of job security: Thomas guaranteed the new job, but he couldn't protect the old one. Also, by then, the harassment had abated, and she felt safe enough to believe that it was over. And besides, Hill was ambitious. Why should a twisted boss stand in the way of her career path?

But after their move to the EEOC, Thomas's tormenting picked up again and became even more personal. He remarked on Hill's body, hair, and whether or not her outfit made her sexy. His behavior as Hill describes it was the very definition of workplace sexual harassment according to the EEOC—Thomas was inhibiting Hill's job performance and creating a hostile work environment. Filing a formal complaint was useless, as Hill now worked for the head of the enforcement agency charged with protecting employees from sexual harassment.

"I felt as though I had been dipped in a vat of scalding water," she said.

The irony deepened when Hill was assigned to review and make recommendations for the agency's sexual harassment policy. Thomas and Hill's colleagues wanted the policy amended to lessen employers' responsibilities in sexual harassment cases. Hill concluded the opposite, that employers should be liable when sexual harassment occurs under their roofs, an opinion she relayed to Thomas. She remembers that he "grumbled and muttered to himself" before begrudgingly accepting her recommendations. Hill wouldn't last in the department long enough to see them implemented.

It is well known that most women who experience sexual harassment at work don't report it. Instead, they're apt to do what Hill herself did: pray that it stops. When it doesn't, plenty of women flee their jobs, their financial security, and sometimes their careers to evade their tormentors. That's what Anita Hill was forced to do in 1983, when she quit working for Clarence Thomas.

It took a medical emergency to propel Hill to seek new work. After a trip to the emergency room, she was diagnosed with stress-related stomach pains. Desperate to get away from Thomas, Hill took the first job offered to her at a small law school in Tulsa, Oklahoma, whose accreditation was hanging by a string. When Thomas badgered Hill to join him for a farewell dinner before she left the EEOC, he warned her that if she ever "told anyone of his behavior that it would ruin his career."

Having been drawn to the pulse and power of Washington,

DC, Hill had never imagined her life taking this path. "I decided to escape the harassment," Hill recalled. "I was settling." She lived off of savings and withdrew retirement money so that she could leave Thomas's office in 1983, and cried the whole plane ride to Tulsa.

Eight years later, in 1991, Supreme Court Justice Thurgood Marshall announced his retirement, and President George H. W. Bush nominated Clarence Thomas to replace him. The NAACP, women's groups, and a consortium of labor interests opposed Thomas's nomination, citing his inexperience and ultraconservatism. He had served as an appeals court judge for just over a year, and they feared he might roll back gains in civil and reproductive rights. Still, Thomas's Senate confirmation hearings consisted mostly of softball questions and evasive answers. He even concealed his views about *Roe v. Wade*, the court decision legalizing abortion, a politically radioactive issue that a high court nominee could never hide from today. His supporters touted his strong character as his preeminent qualification.

Meanwhile, Hill had remade her life teaching law in Oklahoma. After three years at Oral Roberts University, she was hired by the University of Oklahoma. She was still teaching there in July of 1991 when a *New York Times* reporter called to interview Hill for his story about the newest Supreme Court nominee, since she had worked for him. She didn't tell the reporter about the abuse she had suffered, but the traumatic history came flooding back. As she writes in her memoir, she had "for years spent considerable time and effort convincing myself that what happened to me no longer mattered. For the first time I was forced to consider that it *did*—that the behavior was not only an offense to me but unfitting for someone who would sit on the Supreme Court."

Two months later, in September, government officials finally called Hill to discuss rumors that Clarence Thomas had sexually harassed staffers. For the first time, Hill outlined the specifics of what she'd experienced, first in phone calls with Senate investigators and then in a four-page statement. Senate staffers assured her that the statement would remain private. Instead, it was shared with the fourteen members of the Senate Judiciary Committee and their staffs, who were vetting Thomas. (It also launched an FBI investigation, but investigators were inexperienced with sexual harassment and didn't consult experts, leading to an insufficient inquiry.) "I suspected that I would have been treated differently had I had political contacts, money, title, or any other indicia of power," Hill later wrote.

Hill's initial desire for anonymity, or at the very least privacy, was shattered on October 6, 1991, when *Newsday* reporter Timothy Phelps published Hill's allegations. She was immediately flooded with publicity and interview requests. Reporters swarmed Tulsa and ambushed Hill's home, to such an extent that she was forced to move into a hotel to escape them. The Senate Judiciary Committee, which was responsible for vetting Thomas, had practically concluded its confirmation hearings. They did not want to address the sexual harassment accusations against the judge, and led by Senator Joe Biden, the committee fought against hearing Hill.

This rejection stoked Hill's determination to be heard and inspired her to talk to the press. After she spoke to every major television network over two days in early October, the committee caved to the public pressure and agreed to hear Hill's charge on the record. Even then, the senators were more enraged by how the story got "leaked" than they were by allegations that a Supreme Court nominee had sexually harassed a staffer, according to Phelps's book, *Capitol Games*.

"WHOEVER DID THIS OUGHT TO BE SHOT"

Hill had little time to prepare for the hearings—just forty-eight hours to assemble a legal team, find other witnesses, and fly everyone to Washington. But already "the steady campaign to discredit me was in full swing," she remembered. Reports speculated that she was a partisan pawn who had come forward at the last minute to fell Thomas. The committee did not dispel that assumption. Senators who favored Thomas's appointment took to trashing Hill.

"Is this the whole thing, the rantings of a disgruntled employee who has reduced herself to lying?" asked South Carolina senator Strom Thurmond. His colleague threatened violence for accusing Thomas. "Whoever did this ought to be shot," declared Utah senator Orrin Hatch. Hill had wanted to believe in the confirmation process, and thought an invitation to testify before the Senate guaranteed impartiality, but she soon realized how wrong she was.

Hill, her legal team, and her family members arrived at the Senate caucus hall, with its gilded ceiling, marble columns, and lush drapery, on Friday, October 11, 1991. The hearings lasted for three days. The indelible image of Hill sitting alone behind a microphone, testifying opposite fourteen white male senators performing their disbelief on behalf of disbelieving men everywhere, was an education to those who didn't know what sexual harassment was, and a searing reminder of its consequences for women who had experienced it.

The hearing was piped live into living rooms across America. On the first day, Thomas gave a prepared statement denying Hill's charges before she could even make them. That set the tone. After he left the room, Hill entered. She read her own prepared statement detailing her employment history with Thomas, and how he had sexually harassed her and pushed her out of her job in Washington.

Then, the questioning began, first from the committee chairman, Senator Biden. Despite her reluctance, Biden pressed her to detail the "most embarrassing incidents alleged." She talked about the pubic hair on the Coke can, Long Dong Silver, and Thomas's predilection for discussing pornography at work. Senators Arlen Specter and Patrick Leahy harped on discrepancies between Hill's remembrances and her previous statements to the FBI to undermine her story. Senator Alan Simpson called her affidavit "a foul stack of stench."

"Are you a scorned woman?" asked Senator Howell Heflin. He then wondered if she had a "militant attitude" or a "martyr complex," as if only these qualities could explain her charges. Republicans produced an affidavit from Texas attorney and Thomas's law school classmate John N. Doggett III, who claimed that Hill had pursued him romantically, then accused him of "leading her on." To reinforce the scorned-woman theory, Doggett detailed how Hill couldn't stomach men rejecting her.

"It was my opinion . . . that Ms. Hill's fantasies about my sexual interest in her were an indication of the fact that she was having a problem with 'being rejected' by men she was attracted to," he said. Doggett called her accusations against Thomas "another example of her ability to fabricate the idea that someone was interested in her, when, in fact, no such interest existed."

Heflin's "scorned woman" accusation and Doggett's statement contributed to the notion that Hill was an "erotomaniac" or "*Fatal Attraction* type," buzzy terms that crept into the lexicon after the release of a handful of popular movies featuring sex-crazed women villains attacking the men who rejected them or their female competitors. When members of the legal community called Hill's charges a "product of fantasy," what they meant was that Hill had invented the sexual attention that she desired but didn't get from her alleged perpetrators. This depiction boosted Thomas's credibility while crippling Hill's. Polls before the vote to confirm Thomas showed that 55 percent of men and 49 percent of women found him "more believable" than Hill.

EROTOMANIA

The "*Fatal Attraction* type" or the "scorned woman" label stuck to single women, like Hill, who threatened male power. A speedy path to defuse their threat was to assume such women fantasized about sex with men who didn't want them. Erotomania was unsanctioned sexual desire, delusion, and revenge rolled into one. The senators who questioned Hill, and later Monica Lewinsky's most vocal attackers, ensured that these women were stamped with this mark.

They weren't alone. The term was often used in the 90s to smear sex crime victims and to shame female sexuality as predatory and dangerous, particularly when it threatened powerful men's careers. The erotomaniac was devised and summoned by men, reducing female strength and competence to a bodily function. It also bore creepy echoes of Victorian hysteria diagnoses, which targeted and pathologized women's emotional excesses. Erotomania still appears in the most recent *Diagnostic and Statistical Manual of Mental Disorders* (*DSM*). It is categorized under schizophrenia and other psychotic disorders, but "little is known about the background, classification, treatment, or outcome of individuals with this disorder."

Missouri senator John Danforth is to blame for branding Hill as an erotomaniac during her hearings, according to a 2006 paper on "tactics against sexual harassment." The Senate Judiciary Committee member used the "erotomania hypothesis" to prove that Hill was lying. He even sought out an affidavit from a psychiatrist, Park Dietz, who had recently published an article about erotomania, describing it as "a rare delusion of some women that particular men in positions of power . . . have romantic interests in them." Dietz spoke to the press about his theory and how it applied to Hill. The accusation spread like a rash.

In the 90s, erotomania became an increasingly powerful tool to discredit sexual harassment and assault victims, often single women, especially in court. "Date rape" entered the lexicon in 1991, as the rape trial of William Kennedy Smith—the nephew of former president John F. Kennedy—ended in an acquittal after less than an hour of jury deliberation. Considering his high birth, this was "a surprise to no one," quipped Dominick Dunne in *Vanity Fair*. Throughout the trial, defense attorneys and the media assailed plaintiff Patricia Bowman's character, reporting on her drug use, out-of-wedlock child, and abortions. They pinned her as an erotomaniac, a liar, and a floozy who roused suspicion not only for past transgressions but also for being at a bar at three o'clock in the morning, where she met Smith.

Bowman saw where this was headed. "The issue of what I was doing at three in the morning has nothing to do with what happened to me from that man," she said. Three other women gave sworn depositions that Smith had either raped them or tried to, but that evidence didn't make it into the trial. On the stand, Smith testified that his accuser was an "aggressive perpetrator" and an "erratic, hysterical and irrational woman . . . a real nut" who "unbuttoned his pants" and helped him remove her underwear. Like Hill, Bowman was tabbed a "scorned woman" whose accusation of sexual misconduct was recast as revenge. Similarly, Mike Tyson's rape victim, Desiree Washington, was accused of pursuing his fame and money. Vengeance was thought to be the erotomaniac's favorite pastime.

Films dramatizing erotomania helped implant the trope into the cultural narrative. In the late 80s and early 90s, the unmarried woman took on a new, vengeful patina in blockbuster films—she was after your husband for sex, and she'd kill you if you got in her way. In *Fatal Attraction* (1987), Glenn Close becomes so obsessed with her ex-lover that she tries to kill his wife. *The Hand That Rocks the Cradle* (1992) dramatizes the jealous, sex-crazed single woman. When a man commits suicide to skirt assault charges, his widow tries to murder the assault victim. Demi Moore steals an ex-flame's job and sues him for sexual harassment in *Disclosure* (1994). The deranged protagonists in these films took the empowered woman of consumer fantasy, the commercialized "bad girl" who wore makeup shades called Vamp and Vixen, to a murderous and fantastical extreme. Some began to think that the crazed actions of psychotics on-screen could explain real women's behavior. These narratives seemed to warn against remaining uncoupled; singleness could lead a woman to man-stealing and murder.

The erotomaniac label also applied to O. J. Simpson prosecutor Marcia Clark and rock star Courtney Love for their brashness and swagger in male-dominated space. The designation was firmly lodged in the social consciousness by the time Lewinsky came around, and it's no surprise that it was deployed to describe and explain her, too. Tabloids and politicians called her a stalker who must have forced her way into the Oval Office. Clinton aide Sidney Blumenthal testified in the fall of that year that Lewinsky was "a stalker who tried to blackmail Clinton into having sex."

MUD BATH

Hill's defenders thought the Senate hearing resembled a witch hunt—which played out, incidentally, in the very same room where Senator Joseph McCarthy had persecuted alleged communists nearly half a century before. The senators grilled Hill about why she had followed Thomas to another job, why she had kept in touch with him after leaving his employ, and why she had waited so long to come forward to say he had harassed her. Few victims ever do come forward, even today, and when they do they are often punished. Women who accuse can still pay for it. With Hill, it was clear from the questioning that the senators— all white men—did not understand the nature of sexual harassment, nor the experience of being a victim, or even an outsider.

Wyoming senator Alan Simpson produced records of phone calls Hill had made to Thomas's office a handful of times in the nine years since she left his employ, and wondered why she had driven him to the airport after he'd visited the law school where she taught. Hill says she was not threatened by Thomas "as a person" but by "the power he had held over me as an employer." Once she no longer worked for him, the behavior stopped. Thomas was a powerful contact to have in the legal community. "Why should I allow his behavior to deprive me of a job benefit I had rightfully earned?" she asked.

But the committee felt that Hill's continued contact with Thomas must have meant that she couldn't have suffered very much by his conduct. "If what you say this man said to you occurred . . . why in God's name would you ever speak to a man like that the rest of your life?" asked Simpson, the committee member arguably most hostile to Hill. Of course, that she was still infrequently in touch with her harasser—because he was the most powerful person she knew, even years after her departure from his office—did not prove she was unaffected. Rather, it speaks volumes about how his abuse had redirected and gutted her career.

By Saturday's hearing, Republican operatives had unearthed an obscure reference to pubic hair floating in gin in the novel *The Exorcist*, and a mention of Long Dong Silver in a little-known court decision, and funneled these details to Senator Orrin Hatch. He presented these obscurities to Hill, suggesting she had used them to fabricate claims of abuse. The Senate Judiciary Committee lingered on Hatch's accusation and belabored Hill with questions intended to fray her credibility.

"They thought the more they pressed on these details and got Hill to repeat them the more absurd and made up they would seem. I think they were trying to trip her up because they couldn't believe things like this would or could be said," explained

Jill Abramson, the former *New York Times* executive editor, who covered the case as a reporter for the *Wall Street Journal*. But it was exactly these obscene details that made Hill believable, especially to women. Pubic hair on a soda can and Long Dong Silver were anecdotes far too weird for Hill to fabricate.

US senators saying "penis" and "pubic hair" in public may seem quaint compared to the pornified political culture we know now, but it was the first time anything like this had ever happened in Congress—and on television. Reports on Hill's testimony emphasized how her accusations had sullied the hallowed halls of the Senate with sex talk. The press assumed the posture of the modesty police with headlines like "X-rated Drama in a Regal Setting" and "Mud Bath Displaces Decorum." The *Washington Post* called it "the most lurid and dispiriting proceeding of its kind ever in a long history of TV and radio coverage of congressional business," and sympathized with viewers who wanted to "run from the TV set directly to the showers to try to wash the whole thing away."

Nineties television would come to fixate on other stories that preyed on women like the O. J. Simpson trial and the Clinton-Lewinsky scandal. But Hill was first. "It's the first time we've got a story which fascinates the soap opera fans as much as the political junkies," a television executive said at the time.

"THEY ASSUMED THAT BLACK MEN SPOKE FOR US"

Needless to say, race was very much an issue in the hearings. Hill was cast as a promiscuous Jezebel or a frigid, combative Sapphire; Thomas victimized himself, and there was only room for one victim. He finagled speaking both before and after Hill, bookending her claim with his hot denials, calling the hearings "a circus," "a national disgrace," and "a high-tech lynching for uppity blacks who in any way deign to think for themselves." Thomas's race-baiting was even complimented by the White House. Those close to George H. W. Bush said he remained proud of his pick and was "impressed by the way Thomas played the race card." But if Thomas was playing the race card, what did it mean that Hill was also black?

Hill should have understood that Thomas's lewd remarks were flirtations and not complained about them, according to Harvard professor Orlando Patterson. He published an op-ed in the *New York Times* defending Thomas's denial. The piece claimed that Hill's accusations of sexual harassment were derived from a "white, upper-middle-class work world," while Thomas's behavior was akin to courtship in the black working-class milieu, which he claimed to understand as a black man himself. Hill was simply "pretending to be offended."

In response, black feminists fought to dismantle the racist stereotypes at work in the trial, including those invoked by Thomas and Patterson. Elsa Barkley Brown, Deborah King, and Barbara Ransby raised funds to publish a rebuttal advertisement, also in the *New York Times*, titled "African American Women in Defense of Ourselves." In it, they wrote that law and society have always ignored the sexual abuse of black women, while sexually stereotyping them as "immoral, insatiable, perverse, the initiators in all sexual contacts—abusive or otherwise." This was why Anita Hill was not believed, they said. Hill was a victim of intersectional disempowerment, according to more recent writing by race scholar Kimberlé Crenshaw; she was rendered a "white woman" or "raceless" because

Thomas claimed black-victim status with his denial, and because black women's sexuality is alternately demonized and misunderstood. Thus, Hill was the instigator because black female sexuality was a threat. Hill could not be the victim; only Thomas could.

Racism followed Hill even after her testimony concluded. John Burke, who had been a partner at the law firm that employed Hill, submitted an affidavit to the Senate Judiciary Committee describing Hill's work as subpar and stating that he had recommended she find another job. It was later revealed that Hill had hardly worked with Burke, and that her performance reviews were acceptable. Burke had likely confused Hill with another black female associate, with whom he had worked closely.

Hill believes black women's voices are still misconstrued and subjugated by black men's. "Those members of Congress had never even considered that Black women had our own political voice," Hill told Melissa Harris-Perry in an *Essence* magazine interview in 2016. "They assumed that Black men spoke for us." Thus, Thomas's appeals to injustice were heard, while Hill's were not.

Plenty of people believed Hill. To her supporters, Hill's even temperament and intelligent answers during the hearing made the committee's tactics appear even more bullish. "The hearing was so ugly and twisted," former MTV News vice president Michael Alex told me. "When she told that story, it rang so immediately true. You're like, 'Nobody made that up.' Hearing Clarence Thomas's explanation, you sat there and said, 'What bullshit.'" In Alex's newsroom, one woman posted "I Believe Anita Hill" in large letters above her desk. The band Sonic Youth agreed, releasing the song "Youth Against Fascism," with the lyrics "I believe Anita Hill / The judge will rot in hell."

Other women whose testimonies could have changed Thomas's fate have since come forward. His former employee Angela Wright was prepared to testify during Hill's hearing that she had experienced similar treatment by Thomas, but was never called as a witness. Neither was Rose Jourdain, an EEOC speechwriter willing to verify Wright's story. "I knew that Clarence Thomas was capable because he had made similar remarks to me and in my presence about my body and other women's bodies, and he did—he was very egotistical, and he did pressure me to date him, and he did drop by the house when unannounced," Wright later said. She believes that the committee purposely avoided her testimony. A special assistant to Thomas, Sukari Hardnett, said she left the EEOC because she had also witnessed Thomas's inappropriately sexual behavior toward women in the office. "If you were young, black, female and reasonably attractive, you knew full well you were being inspected and auditioned as a female," Hardnett told the *Washington Post*.

THE INTERN

In July 1995, Monica Lewinsky started her internship in the White House. It was the same month that she turned twenty-two years old. In November, she passed the president in the hall and mouthed "Hi," according to Lewinsky's biography, *Monica's Story*, by Andrew Morton. The president smiled back. Lewinsky worked in the same office as Clinton's chief of staff. That same day, the president swung by four or five times

Lewinsky later recalled. He dropped by a staffer's birthday party and smiled at her again. "She was, in White House parlance, getting a lot of presidential 'face time,'" Morton wrote. To up the flirtation ante, Lewinsky flashed the president her thong.

Their affair wasn't a fling. It spanned more than two years, ending in May 1997, and consisted of roughly twenty meetings and "countless" lengthy, late-night phone calls, "all of them made by the President." Clinton promised to protect Lewinsky and told her that she reminded him of his mother. Morton observes: "Far from using her as a mere sexual plaything to be discarded at whim, the fifty-year-old president seemed to have a much deeper need for this girl in her early twenties." And yet America has immortalized the relationship as fellatio in the Oval Office and foreplay with a cigar. In the history books, it isn't the Clinton scandal—it's the Lewinsky scandal. One person is at fault.

In April of 1996, staffers suspicious of Lewinsky's relationship with Clinton transferred her to the Pentagon. Evelyn Lieberman, the deputy White House chief of staff, later told the *New York Times* that Lewinsky was moved because of "immature and inappropriate behavior," and that she was too distracted to get her work done. Lewinsky was devastated by the demotion and what it would mean for her relationship with Clinton. Once at the Pentagon, she befriended a longtime employee, Linda Tripp, and confided in her about the affair. Tripp surreptitiously recorded these confidential talks and then gave the tapes to special prosecutor Kenneth Starr, who was investigating the president in January of 1998. The former Bible salesman seemed to have a bloodthirst for the president and his infidelities.

In love and compelled to protect "Handsome," Lewinsky submitted a false affidavit, professing no sexual relationship with Clinton. But her recorded conversations with Tripp told otherwise. On January 12, Tripp called the Office of the Independent Counsel and said that the president was having an affair with a government employee who was subpoenaed in the Jones case, and that he had told her to lie about it. The next day, Tripp recorded three hours of conversations with Lewinsky.

While many retellings imagine Lewinsky as too dumb to realize her friend was entrapping her, by the end of their relationship, Lewinsky distrusted Tripp and suspected her espionage. She dug in Tripp's purse looking for a recorder after her colleague decamped to the bathroom. She didn't find it because the wire was affixed to Tripp herself. On January 16, armed FBI agents hauled Lewinsky—now twenty-four—to a room at the Ritz-Carlton Hotel and threatened her with charges like perjury, conspiracy, and obstruction of justice, plus twenty-seven years in prison. Agents grilled her for hours. Lewinsky said the encounter felt "as if my stomach had been cut open and someone poured acid onto my wound." She thought about suicide then and there at the hotel, since there were sliding windows she could conceivably jump through.

A CLINTON BIMBO

The scandal, which has been called "the most riveting chapter of recent American history," broke on January 17, 1998, when a blogger named Matt Drudge published a skeleton report that began:

WORLD EXCLUSIVE

MUST CREDIT THE DRUDGE REPORT

At the last minute, at 6 p.m. on Saturday evening, NEWSWEEK magazine killed a story that was destined to shake official Washington to its foundation: A White House intern carried on a sexual affair with the President of the United States!

The *Drudge Report*, which was made by this story, wouldn't out Lewinsky until the following day. She was named in a sexual harassment lawsuit filed by Paula Jones. It took two days for other news outlets to report the story. Drudge's sentence construction cemented the narrative: the intern as subject and actor, the president as object. Early descriptions of Lewinsky noted she was "saucy" and "tearful," and wore "low-cut blouses" and "thigh-high skirts."

After the scandal broke, Lewinsky sequestered herself in her mother's Washington apartment at the Watergate. After she was tailed on her way to get her hair cut, she changed stylists. When she visited her father in Los Angeles, a car full of paparazzi chased her down and crashed into her vehicle. And when the press weren't stalking or endangering her, they insulted everything about her, including her appearance, ambition, and background. Coverage of her focused on tired stereotypes. Lewinsky was slutty, dumb, entitled, and fat. "It's doubtful that Lewinsky played hard to get," *Newsweek* editorialized. Former boyfriends emerged to bash Lewinsky publicly. A married former lover in Oregon called a press conference on his lawn to disparage her.

News anchors wondered if she was "ditzy." Democrats and Republicans alike called her a "Clinton bimbo," a term used to defame the women with whom the president had allegedly cheated on his wife. ABC News promoted their segment featuring Tripp's tapes as the "world's most exclusive girl talk." Of course, Lewinsky and Tripp weren't girls. They were professional women, and assistants to high-level government officials in the Pentagon office of public affairs. But because the topic of conversation was sex with men, the media reduced them to schoolgirls.

FAT AND UNATTRACTIVE

Many blamed Lewinsky for the affair, claiming she wasn't attractive enough to bag the president. "I think what people are outraged about is the way that she looks. . . . I mean, the thing I kept hearing over and over again was Monica Lewinsky's not that pretty," said writer Katie Roiphe at a convening of feminists in February of 1998. Later, when the Starr Report described Clinton complimenting Lewinsky's beauty, sniggers abounded. But more offensive than Lewinsky's looks was her weight. When she was growing up in Beverly Hills, classmates had bullied her and called her "Big Mac."

Lewinsky was self-conscious about her body from an early age. The repeated fat-shaming stung and contributed to "neurotic" behavior regarding her body into adulthood. Tabloids like the *Enquirer* hurled every synonym for "fat" they could find at Lewinsky. They reported on what they thought she ate, from potato chips to crab cakes. Headlines like "Tubby Temptress Checks into Gym to Lose 55 Pounds" were common. The *New York Post* nicknamed her the "Portly Pepperpot." Later, when a famous photographer shot Lewinsky for *Marie Claire UK*, the paper reported that he "overlit her face to make it appear thinner."

It wasn't only the tabloids that taunted Lewinsky for being fat. In the *New York Times*, Maureen Dowd called her "the girl who was too tubby to be in the high school 'in' crowd." Lewinsky had sent care packages to the president containing letters and a tie, but Dowd joked that instead the parcels contained Ho Hos and Ding Dongs (double entendres lost on no one), since both Lewinsky and Clinton were "junk food addicts" likely "swept away on a Slurpee sugar high, comforting each other for their body image disorders." Writing in the *New Yorker*, Jeffrey Toobin described Clinton's young paramour as a budding body-conscious narcissist: "Before she became obsessed with the president of the United States, her only other serious interest in life was dieting," he wrote. Former suitors piled on. Journalist Jake Tapper called her "chubby" and "cute, if a little zaftig" in a *Washington City Paper* article he wrote about dating her. "I have two words for people who think that sex burns calories," cracked Jay Leno. "Monica Lewinsky."

When Lewinsky became a spokesperson for the Jenny Craig diet brand in 2000, some read it as her attempt to trade pounds for love, just as Anna Nicole Smith had hoped to do. Lewinsky had been characterized as "an oversized, oversexed power grubber and home wrecker," wrote Jennifer-Scott Mobley in the book *Fat: Culture and Materiality*. Lewinsky was "performing her weight loss to escape the ghosting of her previous body—and the contempt and disgust associated with it—and thereby re-shaping her public identity," Mobley observed.

But even the "change your body, change your life" fight, which had worked wonders for personalities like Oprah and Kirstie Alley, backfired for Monica Lewinsky. Detractors mocked her for trying to slim down. Jenny Craig reportedly dropped her because she didn't lose enough weight. Her fleshiness rivaled her adultery in its offensiveness.

Not only was she stigmatized for being fat, but Lewinsky's body type contributed to the story line that she was lazy and unkempt—"one of the untidiest people I have ever met," accord-ing to Morton. The emergence of the now infamous blue dress stained with semen underscored her supposed slovenliness. A prurient public had to know: Why didn't she wash it? In her bi-ography, she explained that she didn't want to pay to clean the garment until just before she was going to wear it again, in case she gained weight and couldn't fit into it. She maintains that she told prosecutors about the dress because they threatened her with jail time if she didn't reveal everything, and she had already told Tripp about it on the tapes. But this explanation didn't suf-fice. Instead, it substantiated the many negative traits attributed to her by an unforgiving press and public—gluttony, stupidity, and perversion foremost among them.

HAVING SEX AND LIKING IT

New York Times columnist Maureen Dowd won a Pulitzer Prize "for her fresh and insightful columns on the impact of President Clinton's affair with Monica Lewinsky," judges noted, but the coveted award was won at the expense of a recent college grad who had a consenting relationship with the most powerful man on earth. Dowd's columns about the "ditsy, predatory White House intern" consecrated the notion that she was a dumb slut. She insisted Lewinsky "might have lied under oath for a job at Revlon" and was "immune" from having a brain. Dowd accused her of taking down a powerful man by dispensing her cleavage. A test-balloon defense for Lewinsky could be "the Troubled Slut Defense," Dowd offered. "White House aides note that her friends say Monica arrived in Washington like a heat-seeking missile to seduce the President," she wrote. Lewinsky took to calling her "Moremean Dowdy" among friends.

In the August 1998 column "Monica Gets Her Man," Dowd shames Lewinsky for embracing her sexuality in public. The romance was "pathetically adolescent," and Lewinsky was silly for believing anything to the contrary. "Monica made it clear she was not simply servicing the President. The pleasuring, she insisted, contradicting his account, was mutual. Their relationship was not cheap. It was way unique," Dowd observed.

While slut-shaming Lewinsky was a bandwagon everyone rode, it is striking that this behavior was rampant even within her own camp. Her lawyer and family friend William H. Ginsburg publicly made sexual jokes about Lewinsky. He remarked to her father that Bill Clinton would "cream his pants" if he saw her at a photo shoot. He said that Clinton preferred women with "dark pubic hair," an off-color reference to Lewinsky. *Time* magazine quoted him bitchifying Lewinsky as a "caged dog with her twenty-four-year-old libido."

Fewer things are more threatening to puritanical, morally hysterical America than a young woman having sex and liking it. This is precisely what Lewinsky was guilty of, and the punishment was—and remains—slut-shaming. *Newsweek* reported that Lewinsky's childhood neighbors said she "wore adult makeup" as early as age twelve as evidence that she had long sought male attention. Tapper confirmed this. Though he claimed he was writing about her valiantly, to correct those who called her a "tart," he admitted that he talked to her at a party because she seemed "easy." "I figured that behind her initial aggressiveness lurked an easy, perhaps winning, bit of no-frills hookup," he wrote. On the requisite *Saturday Night Live* sketch, Molly Shannon's Lewinsky insists, "I like BJs," and uses the film *Titanic* as sexual innuendo. "That thing was so long" and "took two hours to go down," she trills.

Like Anita Hill before her, Monica Lewinsky was accused of suffering from erotomania—deluding herself that Clinton wanted her, as Hill had allegedly done with Thomas. The implication was that the affair was her fault and only occurred because her craziness wore the poor man down. Barbara Walters does not hide her disdain for Lewinsky's sexuality, rebuking her throughout the course of their exclusive interview. "The whole country felt you were a stalker and a seductress," said Walters. Critics didn't see Lewinsky expressing her agency by truthfully describing what she knew to be her sexual relationship with the president; they saw her to be threatening him with her lustful story. "Like the Glenn Close character in 'Fatal Attraction,' Monica Lewinsky issued a chilling ultimatum to the man who jilted her: I will not be ignored," Dowd wrote. The stalker sobriquet proved handy. Only a crazy obsessive who didn't wash her clothes after sex could begin and maintain an affair with the leader of the free world, a man whose bedroom door is guarded with guns. The story that women stalked men in large numbers was proved to be a gross exaggeration by a 1998 Justice Department report. It found that women were four times more likely to be victims of stalking than men, and that more than a million women were stalked each year. "Given these findings, stalking should be treated as a legitimate criminal justice and public health concern," the authors wrote.

Lewinsky's self-proclaimed comfort with her own sexuality was discomfiting to many, but women older than her seemed particularly offended. She claimed to come "from a generation where women are sexually supportive of each other," but this concept was lost on older, influential women and feminists, many of whom bad-mouthed her. Betty Friedan publicly called her "some little twerp." Erica Jong cracked that Lewinsky had "third stage gum disease." In her primetime interview, Barbara Walters calls Lewinsky a "big mouth" for telling people about her affair with Clinton. "You're a sensuous, passionate young woman. Is Bill Clinton a sensuous, passionate man?" Walters asks.

"Gosh, I'll probably get in trouble for saying this," Lewinsky replies.

"Not in any more trouble than you're already in, Monica," Walters scolds.

"Where was your self-respect?" she asks, her voice climbing an octave. "Where was your self-esteem?"

"I don't have the feelings of self-worth that a woman should have, and that's hard for me, and I think that's been at the center of a lot of my mistakes and a lot of my pain," Lewinsky answers, making Walters look like a circling vulture, refusing to stand down even though the prey is still very much alive.

Walters asks what Lewinsky will tell her children about the affair. When Lewinsky says she'll tell her children that "Mommy made a big mistake," Walters sums up the interview in a recorded voice-over: "And that is the understatement of the year." Cue the credits.

The Walters sit-down with Lewinsky was the essence of 90s bitchification. Seventy-four million people viewed it, putting it on par with that year's Super Bowl and making it "the most-watched news interview ever televised on a single network."

Sadly, the interview didn't evoke any ire at the older woman's reproachful, finger-wagging treatment of the younger one, or much sympathy for Lewinsky's touching honesty, for that matter. Rather, most calls to ABC after the interview, according to Walters, demanded to know Lewinsky's lipstick shade and where to buy it. Glaze by Club Monaco quickly sold out across the country.

A SHUNNING

Though Anita Hill had charged sexual harassment, after the hearings she was punished for being associated with sex. She was overwhelmed by threats of death, sexual violence, and bombs, and would return home to vile packages and voice mails. "People felt free to leave the most cruel and revolting messages imaginable," she said.

Hill was to appear in a promotional video for her job, at the University of Oklahoma, but after the hearings, the advertisement was pulled and Hill was edited out. "We felt that people would focus on her and not stay with the institutional message," said a spokeswoman. Hill called it "a shunning." Her hometown paper, the *Daily Oklahoman*, ran editorials favoring Thomas's appointment, and called her testimony a ploy conjured by Democratic shills. Hill's advisers discouraged her from responding to the attacks, and few came to her defense.

Meanwhile, Thomas and his wife, Virginia, made the cover of *People* magazine. A photo of the couple hugging opened to a multipage spread featuring them relaxing and reading the Bible in their home. These humanizing images of a married couple seemed to underscore Thomas's innocence. The magazine also ran an essay by Virginia Thomas, in which she accuses Hill of lying and furthers the erotomaniac theory. "What's scary about her allegations is that they remind me of the movie *Fatal Attraction* or, in her case, what I call the fatal assistant. In my heart, I always believed she was probably someone in love with my husband and never got what she wanted," Virginia Thomas wrote.

After the hearings, Hill had myomectomy surgery—doctors removed about eighteen ovarian tumors and cysts to alleviate pain she had suffered from for much of her life, pain that was likely exacerbated by the stress of Thomas and the hearings. She registered at the hospital under a false name to avoid press attention.

Even after the Senate confirmed Thomas's nomination—by the narrowest margin in recent history—the gleeful trashing of Hill continued. David Brock's 1993 book, *The Real Anita Hill*, implies she "effectively committed perjury" during her testimony, which he alleges was "shot through with false, incorrect and misleading statements so much so that . . . it is very difficult to believe what she said about Clarence Thomas is also true." He accuses her of seeking revenge, being opportunistic, and hating men, quoting one of her former law students who called her "militantly anti-male and obsessively concerned with race and gender issues."

While some called the book a character assassination, plenty read it seriously and absorbed its attacks. "It's impossible to finish *The Real Anita Hill* without concluding that Hill failed to be fully honest in her Senate testimony, that she may well have harbored resentment toward Thomas . . . and that she was capable of making obsessive mountains out of ordinary molehills," wrote a reviewer in the *Los Angeles Times*. (Brock later apologized for the book, founded progressive watchdog Media Matters, and became a Hillary Clinton surrogate.)

CLARENCE THOMAS'S PICKUP TECHNIQUE

Saturday Night Live's interpretation of sexual harassment was that it was hysterical. Take the 1991 sketch "Clarence Thomas's Pickup Technique." Joe Biden thanks Anita Hill for discussing "penis size," "the black man's sexual prowess," and "large-breasted women having sex with animals." Then, the senators ask Thomas for sex and dating advice. Adult film actor Long Dong Silver takes the stand.

The sketch celebrates raunch, body parts, and men as dogs, while skewering Hill for being a killjoy. Her abuse accusations are an excuse for the senators to trade crude schoolboy jokes about sex. And Biden gets the biggest cheers of all. To be sure, this is comedy and meant to mock everyone involved. Yael Kohen, author of *We Killed: The Rise of Women in American Comedy*, told me this early 90s period of *SNL* was characterized by "teenage-boy humor." This sketch exemplifies it. But *SNL*'s treatment is how many remember the Anita Hill hearings, and was instrumental in lampooning Hill's legacy.

An *SNL* edition of "Weekend Update" stoops lower. Chris

Rock attacks Hill's appearance, observing that "Clarence Thomas could have picked a much better-looking woman to blow his career on." He then equates harassment with amateurish flirting. "One thing Clarence Thomas is guilty of is using bad pickup lines," he explains. Sex with Hill would have redeemed Thomas in Rock's eyes. "He never ever touched her and he's going to lose the Supreme Court and didn't even get to sleep with her. And that's the real tragedy," he says. Sex with the victim would have bolstered Thomas's reputation with other men, if not with the Senate. *SNL* scoffs at sexual harassment and claims that unattractive women deserve it.

Because she neither slept with Thomas nor projected sexual availability, Hill became the frigid bitch. The all-male Senate Judiciary Committee and the public saw her as "Medusa—the mythological lady with a stare so cold it turns men to stone" and the "vengeful woman of *Fatal Attraction*." The lesson was that men—single or partnered—could have celebrated, prioritized sexuality while single women could not.

YEAR OF THE WOMAN

Today, Hill teaches law at Brandeis University in Waltham, Massachusetts. In her office, a collage bearing the names of scorned biblical women, from Ruth to Jezebel, hangs on her wall. Her career was entirely rerouted on account of Clarence Thomas. She says it became "less about success than survival." Hill's voice would embolden women to share their own sexual harassment stories, inciting marches and protests and eventual policy changes. In the six years following the hearings, sexual harassment claims filed to the EEOC nearly tripled. When President Barack Obama moved to nominate another Supreme Court justice in 2016, following the sudden death of Justice Antonin Scalia, Hill supporters petitioned for it to be her.

It can be tempting to romanticize Hill's story, especially now that her tale is being rediscovered and embraced, vindicating her and enlightening a new generation to the abuse and misogyny she fought against. In the celebrated HBO drama *Confirmation*, Hill's bravery in her darkest days is channeled by the beautiful, talented actress Kerry Washington.

Hill is a true heroine in the crusade against sexual harassment, but she paid a horrendous price. By leveling a serious allegation against a high-profile boss, great forces—the powers that be in Washington, the media, and Hollywood—coalesced to discredit her, defame her, and destroy her reputation and her life. "Being consumed with anger is inconsistent with the goals I have for my life," Hill said on the twentieth anniversary of the hearings. "But of course I'm angry. I'm angry with him, I'm angry with the senators—I'm probably less angry than I was 10 years ago, but it's still there. The larger goal is both gender equality and racial equality, because both racism and sexism contributed to my being victimized."

Sexual harassment and discrimination still plague the workplace. The #MeToo campaign, ignited by the fall of powerful men, like movie titan Harvey Weinstein, amid allegations of rape and assault, revealed just how rampant these practices remain. Sexual harassment continues to threaten women and their livelihoods. Hollywood, politics, media, STEM, food services, law, medicine, academia—no industry or woman is immune. In 2016, anchor Gretchen Carlson settled a sexual harassment suit against the ousted founder of Fox News, Roger Ailes, for $20 million. Carlson said that a reason the kind of harassment she experienced persists today is that companies employed insufficient policies for addressing it in the wake of the Hill hearings. She points to employment contracts that silence accusers as one example. As a result, she encourages women experiencing harassment to hire a lawyer, document it, take the documentation home, and tell two trusted allies.

There are new forums, too, where harassment can be frightening, insidious, and disempowering. Anita Hill's testimony before the Senate predated the internet as we know it. More than 70 percent of adults have witnessed online harassment, while 40 percent have experienced it themselves, according to a 2014 Pew Research Center study.

SINGLENESS: A CAUTIONARY TALE

Those who did pity Lewinsky, or wished to throw her a rope, didn't do so until years later. Once they did, they thought a nice young man her age was just the ticket, that he might lift her from ignominy. "Where is the guy brave enough, strong enough, admirable enough to take her as his wife, to say to the world that he loves this woman even if she will always be an asterisk in American history?" Richard Cohen wrote in the *Washington Post* in 2007. "I hope there is such a guy out there. It would be nice. It would be fair."

The problem with the Clinton-Lewinsky affair wasn't their relationship—the sex, or what it insinuated about abuse of power; it was *them*. Their bodies that ballooned and shrank for all to see. Their rapacious appetites. They succumbed to desires, addictions, out-of-bounds pleasures. Powerful men can fall prey to these things. Women are required to guard against them. In his 2004 book, *The Obesity Myth*, Paul Campos argues that this is why the public became so fascinated with the scandal. Clinton and Lewinsky's failing to be thin and therefore perfect resonated. They may have seemed to be different, special, for having power or hogging headlines, but underneath it all they were just like us. And we were disgusted. The scandal that impeached him, and yet is named for her, is a looking glass. Peer at their missteps, her missteps, and there we are, all of us.

It has become passé to blame women alone for political sex scandals. Former New York congressman Anthony Weiner and former governors Eliot Spitzer and Mark Sanford can attest to this. Weiner sent dick pics to a young love interest while his wife was pregnant. They split because he couldn't stop. Spitzer patronized prostitutes in an upscale Washington hotel. And Mark Sanford vanished with an Argentine mistress, reportedly using tax dollars to do so. Their sexcapades were laughable, and the women they slept with were hardly shamed into infamy. Name one of them. I can't off the top of my head.

Perhaps the most homologous political sex scandal to Clinton-Lewinsky broke in November of 2012, when it was revealed that CIA director David Petraeus and his biographer, Paula Broadwell, had an affair. Reports immediately focused on her clothes, her body, and her off-putting ambition. A general called her "immune to the notion of modesty" while traveling in Afghanistan, and the *Charlotte Observer* reported that she "favored sleeveless outfits that showed off toned, muscular arms." As Jennifer Siebel Newsom, the producer of the media-sexism documentary *Miss Representation*, told me when I wrote about it at the time for the *Daily Beast*, Broadwell was being "Lewinsky-ed."

Lewinsky is a woman who appears to have an exceptionally strong character. She knit scarves for friends while waiting to learn whether or not she would go to prison. She lied to prosecutors not for herself, but to shield the man she loved, and who many believe abused and then discarded her. Through tears, she thanked her own tormentors after they grilled and humiliated her during a merciless grand jury hearing. She feels deeply but does not begrudge her attackers. She is the type who bares her soul and apologizes to those who seem to want to stick her through like a voodoo doll. But these moments of humanity are not how we remember the scandal or the person.

Clinton's sexual dalliances were first whispers, then jokes, and now footnotes. For her part, Lewinsky still struggles to build a public identity untangled from Clinton. Since 1998, she has drifted in and out of the public eye—launching a handbag line, starring in a documentary, moving to London, disappearing for nearly a decade, giving a TED talk, and, most recently, launching a line of anti-bullying emojis. Like Hill, she is spinning her story into a powerful message to help others who suffer from abuse, including on the internet. Sometimes Lewinsky is fully public; other times she's more of a recluse. Who could blame her for being so schizophrenic? After all, we have yet to embrace her story.

Lewinsky became a 90s cautionary tale of singleness. Her sexual confidence and relationship with the most powerful man in the world made her into an erotomaniac and a target of slut-shaming and much worse. She was dehumanized for threatening male power. Two speakers at Lewinsky's alma mater, Lewis & Clark College, in Portland, Oregon, told me separately that they were explicitly forbidden from mentioning her in their remarks, nearly two decades later. At Donald Trump rallies in 2016, vendors sold shirts that read "Hillary Sucks But Not Like Monica." Lewinsky remains a marked woman unable to secure a traditional job that doesn't include the baggage of scandal that has been unfairly named for her, an intern who was seduced by the president of the United States. Nowadays we blame the boss. Perhaps one day students will learn about the Bill Clinton scandal. It would be nice. It would be fair.

4

WOMEN WHO WORKED

The arrest and trial of O. J. Simpson for the murders of his ex-wife Nicole Brown and Ronald Goldman in 1994 and 1995 captivated the nation, and begot an explosive new level of tabloid obsession, the aftershocks of which still pulsate today. Just three years after the Los Angeles riots following the beating of Rodney King by Los Angeles police, the Simpson trial presented the nation with a litmus test on race in America. The daily televised saga also turned on stereotypes of women.

During the trial, Marcia Clark became one of the most famous women in America, and maybe the world. As a prosecutor for the Los Angeles County District Attorney's office, she was tapped to argue the "trial of the century." Clark was not only the lead prosecutor; she was also the only woman. Initially, her boss didn't want her to head the prosecution. He tried to assign her a partner to comanage it, since he wasn't confident she could do it alone. Some speculated she was tapped not for her ability, but because, like the victim, Nicole Brown Simpson, she was a woman.

Twenty years later, Clark is constantly asked to relive a time in her life she describes as "a horrific personal nightmare." She never returned to the courtroom. And with a record like hers—she'd won nineteen of the twenty homicide cases she'd tried by the time she got to Simpson—it was a loss for the justice system. She says she isn't in touch with anyone from those days because, like for PTSD victims, any contact would be too painful a reminder, even decades later.

"What was it that I did personally to make this happen? What do you think I did?" she asks of those who blame her for losing the trial—but she's saying it to me, as we sit in a booth at a Jewish deli in Los Angeles, where I have the salad she usually orders, and she has the chicken soup. "I've never gotten an answer to that question. Not one that makes any sense."

Before Clark starred in the checkout-aisle rags, lawyers and journalists had praised her skills and speculated that she was more likely to win the Simpson case than her opponents. "It is true, too, that Clark is winning because she is every bit the equal of her more celebrated and more highly paid adversaries . . . she may even be better," Jeffery Toobin wrote in the *New Yorker* during the trial.

"She had the energy of a hummingbird," according to her former boss John Lynch. Even her courtroom adversaries agreed. "She gave the most powerful argument I ever heard to a jury," attorney Madelynn Kopple, who lost a double homicide case to Clark, told the *New York Times*. "I was shuddering when I heard it . . . There's nobody tougher than her. Nobody." Clark was eight months pregnant at the time.

"Nobody has a mind like Marcia. She has a fabulous, phenomenal memory of case law, citations, case names," colleague

Susan Gruber shared in the *Washington Post*. NBC senior legal correspondent Cynthia McFadden has covered hundreds of cases in her career, including Clark's prosecution of Robert John Bardo, the stalker and murderer who is serving a life sentence. "And I remember saying to people at the time, 'This is the single best courtroom prosecutor I have ever seen in my life,'" she said.

LAWYERETTE

Clark's perceived competence soon morphed into a threat. A mock jury described Clark as "shifty," "strident," and a "bitch," while categorizing the defense attorney—a man—as "smart." Clark watched the criticism on a video feed in an adjoining conference room. She says that it stung, mostly because it was coming from other women. "The black women in particular viewed me as a bitch in the focus group," she says. "Black women had been some of my best jurors in previous cases."

Being the only woman in courtrooms and at crime scenes full of men was a recurring theme of Clark's legal career from its inception in the Los Angeles District Attorney's office in 1981. Older colleagues called her "lawyerette." She worked in a world where the "standard treatment for a pushy babe is the cold shoulder," she writes in her memoir, *Without a Doubt*. Clark thrived by matching male lawyers' grit. She talked tough, drank scotch, and puffed Dunhills, which inured her to male superiors. Work was her life.

"Trial work is especially appealing to the workaholic," Clark wrote. "I'd go through the docket like Pac-Man, grabbing cases no one else would touch, putting in ten- to twelve-hour days in the process." Eventually, she earned a spot in the elite special trials unit, which investigated and tried the city's highest-profile cases.

Among Clark's accolades as a prosecutor was winning one of the early cases in which DNA analysis, then a new technology, was used to convict a killer even though the victim's body couldn't be found. She also won a murder conviction in the 1991 trial of the stalker who killed the actress Rebecca Schaeffer. The tough-to-get conviction won Clark plaudits from her bosses. "I'm more afraid of ghosts and vampires than I am of killers," she jokes, touching her necklace—a silver bullet dangling from a chain.

SKIRTGATE

In 1995, the same year as the Simpson verdict, the dean of the University of Pennsylvania's Annenberg School for Communication, Kathleen Hall Jamieson, published *Beyond the Double Bind*. In it, she explores the "double bind" that women face in public. "Women who are considered feminine will be judged incompetent, and women who are competent, unfeminine," she wrote. Amy Cuddy, a social psychologist, has also studied this phenomenon and cites how women are penalized when they attempt to exhibit both warmth and competence at once.

Clark was one such pinball dinging between these poles. The initial opprobrium Clark received from the test jurors was soon mirrored out in the general public. Critics called her ruthless, bossy, strident, snippy, chippy, dour, hard-as-nails, flint-hard, a tiger, a miserable wretch, and the chief bitch. Her detractors attacked her femininity, or lack thereof. A comedian recommended that Clark get "mole hair trimmers."

During jury selection in the fall of 1994, focus shifted to Clark's "daring" skirts, which were deemed "too short and too tight." Throughout the trial, her colleagues anonymously pontificated about her "shapely" legs in the press. They were "her one vanity," and earned her the courthouse nickname "Marcia mini." Trial watchers were subject to daily "skirt-alert" analyses from scores of national publications measuring "how much leg the 40-year-old former dancer is showing."

Clark's colleagues in the DA's office weren't the only ones talking out of school about her looks. Simpson's defense attorneys parsed Clark's body as if it were relevant case law. She was "an attractive lady" with "great legs," said defense attorney Robert Shapiro. It's hard to say, in retrospect, if these comments were purely sexist or covertly tactical. He could have meant no harm, but to sexualize her was to puncture her credibility. If she could be reduced to miniskirts, legs, and the mole on her face, surely she couldn't win the trial of the century.

Even the judge presiding over the trial, Lance Ito, leapt into Skirtgate. After a prospective juror told Clark that her skirts were too short, Judge Ito added: "I wondered when someone was going to mention that." Lawyers also condemned Clark in news articles for wearing sexy clothes to sway the court and get her verdict. A tabloid reported that Shapiro "privately expressed some concern that she will use seduction to sway male jurors."

Clark was called "the pinup of the O. J. Simpson trial" and the "prosecutie." A Marcia Clark look-alike in a tiny skirt and fishnets did high kicks on Jay Leno's show. Lawyers and consultants protested Clark's wardrobe, and many advised her to buy more conservative clothes so that she wouldn't jeopardize the state's case. "If it's too short, hang it up," the Associated Press scolded.

The country also fixated on the conspicuous mole above Clark's lip, which "became as famous as Cindy Crawford's," according to the *Baltimore Sun*. A courtroom consultant recommended she accentuate it with pencil, because moles were on trend. The bar's "most memorable mole" also called attention to Clark's facial expressions, which were studied and critiqued mercilessly and said to range from dour to inflamed. Her hair was on trial, too. She had "more than her share of bad hair days," and her perm was nicknamed "poodle 'do." As with many women bitchified in the 90s, Clark-bashing came in Halloween costume form. The formulation was "a mole painted just above the right upper lip, a new hairdo on every doorstep, head held in hands," recommended one newspaper.

While Clark tried to ignore this vapid attention, the sexist critiques lodged in the minds of those watching the trial and helped shape their opinions of the prosecution. Ty West, who coordinated Simpson coverage for *Dateline NBC*, told me, "With Marcia, it was her appearance, what was her hair today, what was she wearing, that kind of thing. It was six paragraphs on Marcia displayed at checkout with a photo. Not a lot of in-depth stuff going on. Very much what people read on the toilet."

One tabloid even ran a topless photo of Clark. It was taken years earlier, during a vacation in Europe with her then husband. Her ex-mother-in-law had sold it to the publication. A secretary in her office tried to comfort her, but she was mortified. "That was a very bad day for me," Clark recalled.

The checkout-aisle treatment undermined Clark's credibility and competence. Her team was often described as "in over their heads." With all the focus on Clark's appearance, to the exclusion of her credentials, it's easy to see why. After the televised nine-month trial ended, Clark told me that people often approached her to say, "I really miss your show."

BITCHY BUT LOVABLE

If CBS had its way, Murphy Brown would have been a young blonde sexpot. Network executives wanted the indefatigable reporter character in the comedy of the same name to be thirty, not forty. Rather than returning to the newsroom from rehab, as she does in the 1988 premiere, they thought she should be back from detoxing at a spa. Producers had hoped to cast the nubile Heather Locklear as Murphy, who would later scorch *Beverly Hills, 90210* spin-off *Melrose Place* as Amanda Woodward, Monday night's notorious man-stealing bitch.

"Heather was very hot then. That was typical of the kind of casting that you would see," *Murphy Brown* creator Diane English told me. English had envisioned Candice Bergen as the unapologetic newshound in the comedy series from the get-go. But the network found Bergen frigid and pretentious. "They felt that she was the ice princess of Park Avenue," English said, and that she lacked "the fire to be this character."

English couldn't have disagreed more with these notes. And thanks to the Writers Guild of America strike, which halted studio productions in 1988, she didn't have to take them. Since she contractually couldn't edit the script, her Murphy Brown—the news anchor who is unmarried, childless by choice, outspoken, powerful, and overcoming addiction—appears in the pilot, which was shot "word for word." English described Murphy as a woman who "lived her life as a man." This Murphy terrified the network, which prized women characters' likability, a word English says makes the hairs on the back of her neck stand on end. But it turned out that English's Murphy was exactly who America had been missing.

While many contemporary comedies turn on the story lines of complex, flawed single women characters—*Broad City*'s Abbi and Ilana, *Veep*'s Selina Meyer, and *Parks and Recreation*'s Leslie Knope—in the 90s this was something altogether new. Murphy Brown was talented and funny, but also selfish and obstinate. At the time, women characters on television typically spoke in honeyed tones and wouldn't harm a bug. That Murphy's character had negative qualities was a revelation.

A crackerjack reporter on a well-regarded nightly news program called *FYI*, Murphy values scoops and lacks any semblance of a personal life. She fights for plum assignments, back-talks to her bosses, and grills interview subjects. Male colleagues cower in her presence. A running joke of the series is her dissatisfaction with incompetent secretaries, whom she constantly fires. She employed ninety-three during the ten-year run of the series. Her suits and shoes are always gorgeous; her hair shines like meringue. Murphy was tough on coworkers but even tougher on herself, making her what critics called a "bitchy but lovable" character. She proved that television was "finally catching on to the reality of many women's lives," according to *The Chicago Sun-Times*.

"I basically wrote Murphy as a man in a skirt," English told me. "I had never seen a strong, competent woman on television who also had the courage of her convictions, who wasn't trying to please everyone, who allowed herself to be rude and who didn't edit herself. These are traits you would normally find in a man." In other words, in order to be taken seriously and push boundaries, Murphy needed to be an aggressor, a woman who would slug someone for calling her a girl. In one episode, she even threatened to kill her more delicate female coworker, mafioso-style.

Fans of all ages loved *Murphy Brown*. The series won eighteen Emmys over ten seasons. Bergen took home five Best Actress awards, which was the most ever given to an actress for portraying a single character until Julia Louis-Dreyfus unseated her in 2017. After winning her fifth in 1995, Bergen declined future Emmy nominations, out of respect for actresses in other shows.

THE RISE AND THREAT OF SINGLENESS

While 90s hit television shows like *Murphy Brown*, *Roseanne*, and *Melrose Place* focused almost completely on white women's struggles at work and in their personal lives, *Living Single* told the stories of black women professionals dating and trying to make it in New York City. Since its inception, television network Fox aimed to attract young and urban audiences with programming that contrasted with the more staid fare on the big three networks—CBS, NBC, and ABC. The television industry began associating Fox with creating "black" hits that, unlike *The Cosby Show*, didn't cross over to attract a diverse viewership, but appealed almost entirely to black viewers who were otherwise faced with a blindingly white primetime television lineup. *Living Single* epitomized this genre. It debuted in 1993, quickly became a hit, and ran until 1998. "Going black is simply smart business," *Newsweek* reported.

The women in the comedy are lively, strong, and drive the action. The show was designed as a vehicle for recording artist

Queen Latifah, who plays *Flavor* magazine editor Khadijah James. She is flanked by her cousin, a divorce attorney, and a retail buyer. Khadijah, Synclaire, Max, and Regine were singular characters on primetime television. Bawdy, entitled, and unapologetically sexual, they were loud and physical with their acting and comedy, and they ribbed one another unsparingly. They were also loyal friends. Watching them parse work dramas or dissect the seemingly endless flow of bad dates, viewers knew that they cared more about one another than they did about men, who were often one-dimensional, naive, or shallow. Male characters—such as handymen, neighbors, and love interests—were most often stock characters, supporting the action or serving as the butt of jokes. This felt like a much-needed role reversal, and a smart commentary on the sitcom genre. When a male neighbor tries to give Synclaire career advice, she listens to her friends instead. A handyman spends an entire episode attempting to reprogram a television remote, and the women point out his ineptitude, telling him they could buy one faster down the block.

The series critiqued the kind of male-centrism and sexism that *Martin* and *In Living Color* doled out. The comedy genre has long satirized injustice and mocked the status quo. On most shows, however, men's stories comprised the meat of the plot, while women were sexual conquests, companions, or mother figures. *Living Single* broke the mold by putting women in the driver's seat and men in the role of sex objects. Soon, the show beat *Martin* in the ratings.

Khadijah's ideal world without men consists of "a bunch of fat, happy women and no crime," she says in the pilot. The women on the show critique men's bodies with lines like, "His butt's so hard you could bounce a quarter off it," reclaiming such gems from the dustbin of male chauvinism. No other show at the time gave women the kind of power that *Living Single* did. And some viewers didn't like it one bit.

The backlash smacked of sexism—and racism. Criticism relegated the female characters to a long-entrenched stereotype: the Sapphire, or the Angry Black Woman. Sociologist Sue Jewell explained that the Sapphire's existence "is predicated on the presence of the corrupt African American male" to "provide her with an opportunity to emasculate him through her use of verbal put downs." Critics' arguments that the characters were "too obsessed with male-bashing" or "viewing men as sex objects" fulfilled this cliché. Despite their college degrees and good jobs, these women appeared to be squandering their achievements with sass and attitude. The black women on *Living Single* were stereotyped as "booty-shaking sugar mamas" and "man-crazed Fly Girls." Such criticisms of female sexuality harkened to the "erotomaniac" insults hurled at Anita Hill and Monica Lewinsky. The insinuation was clear: independent, loud women who excelled at work and controlled their own lives were objectionable—especially if they were black.

Murphy Brown exhibited similar strength to the women in *Living Single*, of course, but she got a pass, probably because she was white. Critics said *Living Single* fell short of its calling, as a classy "black *Designing Women*," because it had "quadruple the sex drive and none of the smarts." *Newsweek* reported that influential black entertainers were irked that Fox shows like *Living Single* cast men as "oversexed, wha's-up, man buffoons." Bill Cosby lambasted the show for limiting its women characters in an interview with the magazine. "Suppose I did a sitcom about four African-American women like Fox's *Living Single*? In my show, two of them might be sitting around discussing men all the time. But the other two women are going someplace, something else is happening in their lives. Is that too much to ask?" Cosby's subsequent credibility implosion aside, he ignores the many episodes in which the women on *Living Single* struggle with identity and careers. Synclaire fights for a promotion. The magazine she works for in the show reports a story on embezzlement at city hall. They play Scrabble. They also smartly satirize the crap black women put up with from black men, according to the show's boosters.

The creator of *Living Single*, Yvette Lee Bowser, resisted a show that revolved around sex and dating. In a 1997 interview, she spoke about kowtowing to network pressure to strengthen the male characters and gear the show's plot toward the women dating them. Fox reportedly told her, "You're not going to get your female show on the air without a strong male presence." So she acquiesced.

After the show had been called man-hating and too black, and its female characters dumb and frivolous, a curious thing happened. In 1994, a year after *Living Single* first debuted, a new sitcom hit the airwaves. Like *Living Single*, this new show also starred young urban singles. Unlike *Living Single*, it cast three women and three men, all white. Industry critics quickly recognized that NBC had copied *Living Single*'s winning formula and pumped it full of cash to make its own hit show, *Friends*. Perhaps *Friends* made itself more palatable by amping up the male characters' dilemmas and dialogue, and tamping down the female characters, who were the strong center of *Living Single*. In *Friends*, the women fret about sex and dating and are often packaged into stereotypes, like Monica (the former fat girl and control freak), Rachel (the daddy's girl), and Phoebe (the crazy one). They work domestic-style jobs in food services or fashion. Unlike Khadijah and Synclaire, they are mostly deferential to their male foils. *Newsweek* compared the bodies of both casts. *Friends'* female stars were "classically beautiful and reed thin," while *Living Single*'s were "a less Hollywood ideal" and "black women whose bodies are, well, real."

The reedy, milquetoast gals of *Friends* signaled the beginning of the end for the fiery, independent women of *Living Single*. Rather than point out the debt *Friends* clearly owed to *Living Single*, and how *Living Single* created new possibilities for women on television, coverage of the cast's reaction to *Friends* stoked a catfight. Queen Latifah was reportedly annoyed by the new series's fancy billboard erected outside the Warner Bros. lot where both comedies filmed.

While the *Friends* cast would "cavort each week to the sound of their hit theme song," *Living Single*'s was now "singing a much different tune," according to a *Los Angeles Times* report called "'Single' Looks for a Little Help Against 'Friends.'" Bowser lamented that "we have never gotten that kind of push that 'Friends' has had." She aired her grievances with Fox for stifling her creativity and providing only lackluster promotion of *Living Single*. After *Friends* debuted, *Living Single* hung on for several more seasons before being canceled.

CATCH A HUSBAND

Living Single offered an independent ethos during a time of marriage decline. Between 1990 and 1999, the median marriage age rose from twenty-three to twenty-five, a significant increase considering it had previously hovered between twenty and twenty-two for almost one hundred years. More women were staying single longer, or skipping nuptials altogether. This signaled a national shift in priorities, away from traditional gender roles and toward increased sexual and financial freedom and a more independent existence for women.

The shift also triggered a vicious cultural backlash that warned women against uncommitted sexual agency and pressured them to marry. The counternarrative—fueled by the epidemics of AIDS, STDs, and date rape in the 90s—insisted that women didn't want sexual freedom after all. It was too dangerous. What women wanted, and had wanted all along, was simple, safe monogamy—sex within the confines of marriage.

"The dominant impression in the media was that a single woman now wanted 'commitment,' not carnality; 'courtship,' not casual sex," wrote Wendy Dennis in her 1992 study of sex, *Hot and Bothered: Men and Women, Sex and Love in the 90s.*

Well-timed dating guides stepped in to exploit the anxieties and confusion about women's shifting place in society. Dating self-help hit new heights, largely propelled by two seminal books that promised women the security of commitment in a handful of steps. *The Rules* and *Men Are from Mars, Women Are from Venus* seemed to create problems between men and women, then propose to fix them. They offered women the alluring prospect of everlasting love in return for retrograde self-sacrifice.

Published in 1995, *The Rules* guaranteed women happiness through sublimation. The tactics of the "beribboned husband-catching primer" included never approaching men, waiting to be approached, and never accepting a date for the weekend after the middle of the week. The book forbids women from moving in with men before marriage, and tells them to dump suitors who don't buy romantic birthday gifts. Other advice included drastic diets, grooming regimes, and plastic surgery. Women shouldn't let a flat chest or large nose inhibit their eligibility, authors Sherrie Schneider and Ellen Fein reasoned. "If you have a bad nose, get a nose job," they recommended. Lots of women in the 90s seemed to comply; between 1992 and 2004, breast augmentations increased more than 700 percent.

The Rules marketed itself to women who had been burned under the mantle of feminism, women who had achieved careers but lacked romance. This characterization at once celebrates and shames women's workplace successes. "Women want to get married, and the way they've been acting for the last twenty years hasn't worked," Schneider said at a seminar. Being bosses had fulfilled women but stymied their love lives. Now they risked becoming the spinsters of American nightmares. They were to blame, but now there was help.

Boosters believed *The Rules* offered women agency. The text lists tools to "take control" of relationships and discourages women from "throwing themselves at guys who aren't interested," wrote a former adherent. It smashed bestseller lists. A cult following mushroomed. The authors became instant relationship experts and pseudo celebrities, ministering to lonely-hearted fangirls at workshops across the country.

But criticism mounted that these rules were outmoded and required women to return to obedient, deferential femininity. Reviewers disparaged the book as your mother's dating advice, asking readers to "forget equality" and "revert back to the good ol' days of playing hard-to-get and easy-to-be-with." Many wondered whether it was "another manifestation of the New Conservatism," a wild-female-taming plot authored by Newt Gingrich and the Moral Majority.

Rules girls, as adherents were called, were labeled helpless, desperate spinsters who couldn't land men. *Dateline* did a segment on the book that took viewers inside one of the conference rooms where women paid forty-five dollars each "to erase years of bad dating habits." Correspondent Josh Mankiewicz explained that "Understandably, most of them wanted to stay anonymous," spotlighting their desperation.

Male pushback to *The Rules* was fierce. "This rules thing may be about the *M* word, but it doesn't stand for marriage," *Dateline*'s Mankiewicz says slowly, teeing up his big reveal. "What we're talking about here is manipulating men, aren't we?"

"You'll never believe it's training. You're going to enjoy it so much," Fein says.

Critics charged the authors with using guerrilla tactics to ensnare men. "Thousands of unsuspecting American males are, in theory, being lured, hooked and reeled in by legions of determined women using military-style tactics," poked one paper. Mankiewicz called their conferences "boot camp . . . with the drill sergeants of tough love."

Schneider and Fein couldn't have predicted that wounded men would devise ways to beat them at their own game. A trio of childhood friends said *The Rules* were wrong and decided to write the antidote, *What Men Want*. In it, they ask that women cook them dinner and stroke their fragile egos. They distribute women into two categories, "wife potential" and "good for now." "If you have sex with him too soon, he will be less likely to consider you as a potential girlfriend or wife," the men write, adding that while men may behave like dogs, "deep down they are conservative and idealistic about the kind of girl they will marry."

Psychologist Meg-John Barker argues that the anachronistic dating advice and popularity of *The Rules* in the 1990s roused the alpha male seduction communities of the 2000s, which flourish today. The pickup artist movement—PUA for short—taught men to seduce women for sport using chauvinistic tactics. These were codified in Neil Strauss's 2005 book *The Game*. Predatory pickup artists were seen as a direct "response to hard-to-get femininity," especially in a culture inclined to blame women for men's actions.

As dating began to move online in the late 90s and early 2000s, *The Rules* pivoted to help women meet men there. Matt Lauer interviewed *The Rules* authors on the *Today Show* about their new online dating guide in 2002. He seemed suspicious of online dating, and suggested that women who advertise themselves for dates were probably slutty.

"What kind of online profile lands a Rules Girl a man?" Lauer asks.

Schneider describes a success story. "Brunette Beauty was her screen name, which was clever and visual and not 'Live It Up' or 'Looking For My Soul Mate 51,'" she says. "Loose and Easy," Lauer jokes, offering an example of his idea of a bad profile name. "Yeah. Right?"

"Nothing about, 'I've been hurt before, I hope you're different,'" Schneider cautions. "Because then people think: damaged goods," Lauer says, throwing up some air quotes.

While single women were seen as either desperate or slutty, the 1992 dating guide *Men Are from Mars, Women Are from Venus* prioritized male sexuality under the auspices of supporting each gender's inborn nature. Author John Gray advised women to capitulate to men to get dates, no decoy coating of empowerment necessary. By 1997, Gray's *Mars/Venus* had sold six million copies, was published in thirty-eight languages, and earned Gray an estimated $18 million. Not only were single women and men not on an even playing field, they weren't even on the same planet.

This polarization of the sexes, and the prioritization of men's needs featured in *Mars/Venus* and *The Rules*, both reflected and flowed into the culture. Followers internalized the idea that gender transcended shared humanity, and that women were a subspecies, especially in the bedroom. This mind-set erased nonheterosexual experience, while mandating that the only way to experience sex and love was through outmoded gender roles. Most of all, *Mars/Venus* and *The Rules* propped up the myth that single, working women should ditch their desires and agency to pursue monogamous partnerships, and that their worth lay in whether they appealed to men.

PASTELIZE

After the encounter with the mock jury, advisers recommended that Marcia Clark tone down her brassy image to appeal to those who called her a bitch. Jury consultants suggested she "dress softer, speak more softly, don't be strident," Clark recalls with a laugh. "I don't blame them for saying, 'Try to soften up a bit and make yourself more relatable, more palatable,'" Clark says. "So they're going to talk about my hair and my skirts? Fuck it. Whatever. I don't have time for this. The clothing was nothing really. But I did wear longer skirts. I did pastelize myself, you know? I went along with it because I didn't care. It wasn't that important to me. I'm not trying to make any statement with my clothes. So if my clothes are getting in the way, then let's just take that off the table. That was the least of my problems. I had bigger fish to fry." Trial lawyers often use consultants who recommend tweaks to make them more appealing to juries. But the recommendations for Clark to soften were inherently sexist.

While Simpson defense attorney Johnnie Cochran's suits were noted for their flash and ostentatiousness, far less ink was spilled on the subject of his looks than Clark's. What's notable is how often Clark's appearance was compared to Cochran's courtroom performance, to prove he was the better lawyer. A *Washington Post* profile describes her as "looking peacockish" for talking to her colleague. "For all his preening, it is a mistake Cochran never makes," the paper observed. A retired FBI agent whose specialty was using psychology to catch serial killers told *Salon* that Clark was "outclassed" by Cochran. "She was just flying around like a little bumblebee as he was spinning the web and she just flew right into it," he said.

Any small change Clark made to her appearance was noted and criticized. When she altered her hair, she made the news. "We didn't have dry shampoo!" she laments. "In the middle of the trial my perm is falling out. I didn't have the time or the money to deal. My hairdresser was telling me, 'Don't perm it, it's drying out and it looks like shit.' So I let it go. One morning I blew it out. I let it loose." When Clark walked in with straight hair, the courtroom tittered. Individuals jokingly wondered whether Clark was present at all, because she looked so different. Ultimately, they gave her a standing ovation, with press kvelling over the "prosecutie's sexy new look." They described it as "softer" and "more feminine." When she began to wear concealer to cover the bags beneath her eyes, it was reported that she had had a makeover. An uncharitable report said she "needed a make-over of her makeover."

Speculation mounted that Clark's "new look" had come from a sinister place. It was inauthentic, critics charged, and a ploy to manipulate the public who so doggedly watched her every move. This puzzled Clark, who thought the focus on her appearance was "stupid and ridiculous."

"Only Hillary Clinton has gone through more repackaging for a public that still hasn't decided if it wants women to work, let alone be good at their jobs," observed the *Baltimore Sun* in an article about reactions to Clark's makeover.

With such tremendous attention trained on Clark's looks, it was natural that they would be compared to other women's looks, and that she would be pitted against other women who had nothing to do with the trial. The *San Francisco Chronicle* polled women trial lawyers, not on their opinions of legal strategy, but on what they thought of Clark's makeover. They were not kind. In fact, they were "troubled" because they felt Clark had cheated to get to the top. They had "managed to succeed without a cosmetic make-over." "Marcia Clark's New Look Irks Female Lawyers," trolled the headline, promising a catfight.

Other reports charged that Clark's persona had changed with her hair and makeup—once too angry and harsh, she was now giggly and sweet, joking around during frequent sidebars and escalating the drama for the cameras. But Clark challenges the notion that she became a pushover: "The problem then becomes, OK, so I go in and wear a pinafore and curtsy. OK, you sent a cream puff in to do a real person's job. I'm not issuing

party invitations, kids. This is a prosecution of a double homicide. I have to duke it out with the defense team. I don't see how
I can be sweet." According to defense attorney Alan Dershowitz,
who was on deck to handle an appeal, the "shrill," "unprofes-
sional" Clark "cried wolf" so frequently in court that by the end
of the trial, no one took her seriously anymore. Outsiders watching the trial noticed how the defense attorneys and judge inter-
rupted her and talked down to her, and wrote off her responses
as histrionics.

Clark contends that in most office environments, people
behave like themselves, more or less, but that's not true of court.
"The drama's inherent in the job. You know somebody killed
somebody. You're prosecuting them. They're on trial for their
life. It's a dramatic situation," Clark says. "Of course I'm different in court than I am outside. Everybody was like that. I
wasn't unique in that way. Court is not your average workplace.
It's formal. It's dramatic. It's war. People go into court and they
are different because it's a trial and you're at war. It took me by
surprise that people were surprised by that." Yet this drama, inherent in any murder trial, and perhaps more so in a televised,
celebrity-laden one, was thought to be Clark's personal province.

Women's groups flagged Clark's mistreatment in court as
obviously gendered. Clark says that Judge Ito would address the
lawyers, "Mr. Cochran, Mr. Shapiro, Mr. Darden, and Marcia."
The Los Angeles Chapter of the National Organization for
Women accused Judge Ito of taking Clark "less seriously than
other attorneys," in a formal letter charging sexism, and explaining its potential to shift the case's outcome.

"We are concerned however with the impact your perceived
attitude may have on the jury. After all, if Your Honor appears

not to take Ms. Clark as seriously as Mr. Cochran, why should they?" wrote Tammy Bruce, president of NOW's Los Angeles chapter. Clark told me that she was surprised by Bruce's accusations at the time and wasn't sure she agreed. "I think I did notice he was harder on me than the other lawyers, but I didn't make the connection to it as a 'woman thing' as much as I figured it was because he was such a celebrity whore. I do admit that I have a tendency to overlook sexism. I ignore it or laugh it off, just push it away. Especially if calling someone out on it may cause harm in the context of a trial, piss off a judge, and wind up with lousy rulings," she says.

Bruce believes her letter and subsequent meeting with Ito to discuss the charge—to which she brought video clips of specific incidents—forced Ito to treat Clark better in the professional setting and camera glare. Clark agrees. "For the next month, he actually treated me like a person," she told me. But this wouldn't last.

"CRAZED HORNY BITCH"

Looking back, it's hard not to see how gender also tinged Lewinsky's ambition and made it laughable. Her desire to succeed at work was considered evidence of her striving and sluttiness. Many believed Lewinsky was entitled and assumed she had demanded that Clinton give her a job. Barbara Walters, in her primetime interview with Lewinsky, hit this like a bug that wouldn't die. "You know other women have had sexual relationships and they don't expect that out of it is going to come a job," she said. "I never expected that out of this relationship will come a job," Lewinsky said. In fact, she had lost her job at the White

House, and had been demoted to the Pentagon in retaliation for her relationship with Clinton. She had worked in the White House before the affair. She hadn't demanded a perk; rather, she pushed to reclaim the job that was taken from her. Walters took her harping as proof that "toward the end of this relationship, you were a real pain in the butt." Morton, Lewinsky's biographer, wrote that "the greatest irony of all" was that Lewinsky was accused of "getting a job as a consequence of the very relationship which had blighted her career." Lewinsky's ambition was used against her. She couldn't be both feminine (read: sexual) and competent. She appeared to want—and to take—too much.

Just as Lewinsky's and Hill's attackers had done, Clark's foes worked to shred her credibility and accomplishments by accusing her of being a sex fiend who refused rejection. Defense attorney (and her adversary) Robert Shapiro claimed he had rejected her advances. "My unwillingness to flirt with her . . . made her angry," he boasted in his memoir, *The Search for Justice*. In contrast, Johnnie Cochran had a great relationship with Clark, Shapiro wrote, because they flirted with each other. He added that Cochran's wife objected to her husband's tactic, feeding catfight fantasies. Defense attorneys "found her a 'whining' minx with an uplifted nostril and a rehearsed hurt look," reported the *Washington Post*.

"Flirting? Seriously? Because I smiled at him? I mean, come on. Either I'm too stern or too playful. It's a lose-lose proposition," Clark says.

Clark was even accused of using sex to puncture Johnnie Cochran's game. According to Dershowitz, Clark allegedly told Cochran, "When you're up there, I want you to think of only one thing: I'm not wearing any underwear."

"She uses all of her resources. You got to give her credit," Dershowitz told CNN anchor Martin Savidge, to which he replied, "she used every opportunity she could."

"There was no underwear comment," Clark told me. "That never happened."

Clark's love life was an obsession during the trial. Aside from her alleged flirtations with her opponents, the rumor mill romantically linked Clark to her coprosecutor, Christopher Darden. In a *Saturday Night Live* sketch, Darden (Tim Meadows) and Clark (Nancy Walls) pursue a romantic relationship in flashback while Barbara Walters (Cheri Oteri) interviews Darden in the present, satirizing the bloated media coverage of the trial's intricacies. Darden tells Walters in an interview that Clark forced him into bed with her.

"Not all of us are racist, Chris," Clark says. "Some of us think black is beautiful." She rubs Darden's shoulders and aggressively kisses his neck as hip-hop begins to play.

"Marcia, you acting crazy," Darden, the voice of sexless reason, says. "We got to stay focused on the trial. This is wrong." She straddles him despite his pleading, "No, no. Help me!" Clark ignores his distress. After the not-guilty verdict is announced Darden sulks but Clark remains oblivious.

"The only thing I'm guilty of is being extremely horny. Please remove your pants," Clark says.

"Marcia, we just lost the case of the century. O. J. Simpson got away," Darden protests.

"Quit whining, Chris. It's time to take the black Bronco down the 405."

"I'm so upset, I don't think I can do this," Darden says.

"Certainly a very different Marcia Clark than the one we're

accustomed to seeing," chimes in Walters. "Smelling the same musky man scent day in and day out could certainly turn any woman into a crazed horny bitch."

Clark's character on *SNL* is built on male fantasy, it goes without saying. But it's worth examining the nature of the humor, which plays on the belief that powerful women are sex-crazed beneath their suits, so much so that the suggestion of a man's attention renders them predators. Watching these decades-old clips, it's notable how even-tempered the men in these scenarios are. If a woman appears too strong to be victimized with sex, she is a secretly sex-starved carnivore, an archetype to be shamed and feared.

Erotomania, like the kind *SNL* imagined Clark suffered from, "is a projection of men's preoccupation with sex," said linguistics professor Deborah Tannen in the *New York Times*. Author of the bestseller *You Just Don't Understand*, about miscommunication between men and women, Tannen calls erotomania what it is: a male creation. "Men think about sex all the time, so they want to believe that women do, too."

THE DOUBLE BIND

When Clark tried to do her job as a prosecutor, she was often accused of being harsh, strident, and unlikable. When she softened her image with pastels and giggles, detractors besmirched her professional credentials and called her "a flirt in lawyer's clothing." What was happening to Clark seems clear now, with more than two decades of distance, but the efforts to discredit her as incompetent or unfeminine were so insidious that even women watching the trial closely didn't see them at the time.

Clark has achieved quite a lot since the Simpson trial. She has authored six books and consulted on television shows, and she reviews appeals cases for indigent defendants, but people don't know her for these things. She is forever famous for what is perceived to be her biggest failure: the inability to convict O. J. Simpson of murder. After the trial, Clark briefly became a legal correspondent, talking about court cases on TV. People cracked jokes that she wouldn't just describe the cases, but also what she would do to lose them.

The day of the verdict, Clark walked out of her office for good. Someone else cleaned out her things because she couldn't bear to look at them. The case ended her courtroom legal career. Like many of her colleagues from the trial, she won a lucrative book deal, but was retraumatized writing about her experience during the Simpson trial.

Clark said penning *Without a Doubt* "was like reliving the nightmare and dissecting the nightmare. It was just terrible." Few have connected the media's and Ito's skewering of Clark's competence to the trial's verdict—O. J. Simpson's acquittal. It's a thread long buried beneath countless other theories of the case.

More than twenty years after the verdict, in late May 2016, Marcia Clark appears at the 92nd Street Y in New York City to promote her new crime novel, *Blood Defense*. The "semi-autobiographical" novel follows an ambitious female defense attorney, and is being adapted into a television series. On a stage beneath a decorative proscenium, Clark sits opposite NBC senior legal correspondent Cynthia McFadden, who asks her to rehash trial gossip and reveal what it was like to watch a Hollywood actress play her on television.

There are about three hundred people in the audience. The

man seated next to me followed the trial closely. He asks me if that's what her book is about. The couple behind us wonders the same thing. He points out the *60 Minutes* journalist Lesley Stahl in front of us, turning away an autograph seeker with a flick of her hand.

McFadden covered the Simpson trial in real time and spent the 134 televised days of it in spitting distance of Clark. "But we never, in all that time, talked," she tells the audience. "Marcia Clark was the prosecutor intent on keeping her dignity in the courtroom and her mouth shut out of it."

Clark is back in the spotlight during and after the conclusion of the TV series that brought new and old audiences to the Simpson-Brown story. She watched her personal nightmare play out on FX week after week for more than two months, dramatized and rendered in soft focus by Hollywood. "It's painful to watch," she tells McFadden. "Maybe if I was over it, it wouldn't be."

With the launch of the FX series, Clark has given countless interviews and appeared on *Ellen* and *The View.* Everyone wants to discuss, for the first time, the misogyny she experienced, and how they didn't see it then, but so clearly do now. *The Cut* proclaimed that Clark is "set to become a feminist icon."

"The way you were treated in comparison to the way the male lawyers were treated," McFadden says. "I was shocked at myself that I was not more aware of this at the time."

"No one really talked about it," Clark says. "And neither did we. Women in the 90s doing a man's job had to be tough. And if you call sexism, you're a lame excuse for a whiner. It's all about your weakness. You can't take it. So I never complained."

Clark and the actress Sarah Paulson, who played her in the series, became fast friends during the show's taping. They even text each other Bitmojis while Clark and I are having lunch. Later, she'll attend the Emmys as Paulson's date and watch Paulson win a best actress award for portraying her.

While Clark's feminist redemption seems in full force, there is also a fresh round of criticism. When NPR's Terry Gross interviewed Paulson about what it was like to play Clark, the actress recalled the journalist Jeffery Toobin—whose book *The Run of His Life* was the basis of the new series—sharing his views of Clark on set. "I like him very much, I just don't agree with some of his opinions about Marcia," Paulson told Gross. "He said something to me in person about her arrogance or something and I just excused myself politely and went back to my seat and mumbled something to Sterling, who plays Darden. I was like, 'Ugh. I don't want to talk to him if he's going to talk to me that way about this person that I'm playing that I've come to revere and feel so much empathy for and compassion for. Also he's a man. What does he know? Get out of here.'"

When I ask Clark how this new fandom feels, she assures me that she wasn't anticipating it. "I never expected anyone to pull out the rampant sexism in the trial," she says. "I just hope this new awareness continues so it will help all women."

ADORABLE DOPES

By the late 90s, as *Living Single* was canceled and *Murphy Brown* was signing off for good, it seemed like strong working women on television might be a short-lived phenomenon. The educated, distinguished career woman was becoming absent from the airwaves unless she was "sexless, old and a bitch," according to one talent agent quoted by the *Chicago Tribune*. Instead, television favored a new working woman: the "Adorable Dope," as coined by *That '70s Show* creator Bonnie Turner. A fleet of women characters popped up in sitcoms and dramedies who were dismissed as ditzy, dopey, hapless, daffy, and easy prey. Journalist Nancy Hass described the Adorable Dope in the *New York Times* in 1998: "She is young, perennially confused, perpetually underemployed and adorably confounded by men. In her teens and early 20s, she is smart and spunky, but approaching 30 she is mysteriously stricken with an unnamed disease that renders her increasingly incompetent. Miraculously, her cuteness is left intact."

The Adorable Dope was a "dizzy girl-woman who represents an updated, politically correct version of yesterday's bimbo," explained the *Los Angeles Times*. And on late 90s television she was everywhere. Because she was hyperfeminine, her ineptitude was darling. She needed men to make decisions for her. Adorable Dopes were known for "having hearts of gold and brains of mush." And, perhaps most loathsome was their "scarfing down low-fat Doritos."

Adorable Dopes held jobs specifically sanctioned for women—waitress, sex writer, head of a panty company—and their careers were often a sideshow to their love lives. Rachel Green of *Friends* is the archetype. She struggles to cut up the credit card her daddy underwrites, and can't, as a waitress, fulfill a simple coffee order. Other Adorable Dopes were powerful on paper—executives, business owners, and attorneys—but they were all hapless when it came to life's basics. Kirstie Alley's Veronica in *Veronica's Closet* is a lingerie CEO who can't run her own life, let alone a global business. Sarah Jessica Parker's Carrie Bradshaw in *Sex and the City* pioneered a certain type of third-wave feminist antihero and reinvented a struggling HBO, but was scoffed at for her sex column, romanticism, and shopping addiction. Jenna Elfman's Dharma Finkelstein, the free-spirited child of hippies in *Dharma & Greg*, taught yoga and trained dogs, but was mocked for being too nice. Calista Flockhart's Ally McBeal was an Ivy League-educated lawyer, but she was an Adorable Dope nevertheless because of her coquettish demeanor in the office and her incessant dating and personal problems.

No one claimed that lingerie company CEO Veronica in the NBC series *Veronica's Closet* was a feminist role model. Alley's character is bossed by underlings and chauffeured by her father. She marries a creepy stranger while she is blackout drunk in Atlantic City. "Her indecision and compelling need to please others are curious personality traits for the chief executive of a multimillion-dollar lingerie company," observed San Diego State University professor Martha Lauzen in the 1999 *Los Angeles Times* article "Alpha Females Still Trail Adorable Dopes."

Veronica's Closet was one of the only shows on air at the time with a female lead over forty years old, which is still a rarity in primetime. Turn thirty-nine and women "simply drop off the primetime planet," according to Lauzen, who found that thirty-something women made up 40 percent of all characters on television. Women in their forties made up a paltry 15 percent. Though Veronica was forty, she acted like a child. In one scene, for example, she prefers to be dragged on the floor with her deceased dog's leash rather than part with it. When Veronica lays the physical comedy on thick, critics accused her of seeming "to have lost both her mind and much of her simple motor coordination."

SHE KILLED FEMINISM

Probably the most infamous Adorable Dope was the one who was presented as a bellwether of the modern woman, but was then accused of killing feminism when she wasn't progressive enough for viewers. Because Ally McBeal had the Harvard law degree, the sparkling career, and the freedom of no spouse or kids, women had high hopes for her. Show creators presented her as a reflection of the modern, ambitious career woman delaying marriage by choice in the 90s. "If women wanted to change society, they could do it," Ally says in the show's 1997 pilot. "I plan to change it. I'd just like to get married first."

Ally quickly attracted the dope label because she had the "emotional klutziness of a teenager and the same level of self-involvement," snarked the *Chicago Tribune*. She won in court, but was "regularly defeated by her neuroses," wrote the *Denver Post*. The "lawyer in a miniskirt" was easily felled by a cursory look from her ex-paramour, and constantly mothered by her wiser, more womanly roommate. She hallucinated all manner of things, including dancing babies she'd never have. Feminist cultural critic Jennifer Pozner lamented that such a "wimp" was positioned to speak for feminism, and called her "a shallow, bratty, willful adolescent with an adult woman's career, a supermodel's miniskirts, and a high school girl's dating anxieties." Flockhart's shrinking figure and assumed eating disorder didn't help matters. She was labeled a "postfeminist icon" representing the "me generation" of women taking their privilege for granted.

The *Denver Post* reported that Flockhart was "frustrated by feminist arguments that her character is ditzy and a poor role model for professional women." In an interview, Flockhart said of Ally, "I think she's got flaws and she's certainly not politically correct all the time. That to me is interesting." She points out that her male costars hadn't become role models for men, adding, "But people put Ally in this bracket of 'she must be a role model, she must be a good one.' It's not really fair."

When Ally wasn't dopey, she was jealous and catty, calling a female courtroom opponent a bitch because she is pretty. She even beat up her crush's wife in kickboxing class. Ally was often a walking catfight waiting to happen.

The agita surrounding Ally came to a head when she appeared on a June 1998 cover of *Time* magazine along with images of iconic feminists such as Susan B. Anthony, Betty Friedan, and Gloria Steinem. Beneath Ally's head, in red letters, the cover asked, "Is Feminism Dead?" According to the magazine, the answer was yes; short-skirted Ally McBeal had killed it. Women had gone from "bra burning to ohmigosh, I just wanna have sex!" as the *Orlando Sentinel* mockingly put it.

The problem with Ally, it seems, is that she was something new for television viewers. She was extremely flawed and neurotic, but also likable and relatable, and often fun to watch, even now. Her downfall was that she was hoisted up as an icon for progressive women to embrace, while not being progressive enough for some women and being far too progressive for many men. Television offered so few complex women characters for women viewers (its main audience, by the way) to savor, so it was painfully easy to pick apart a character like Ally McBeal, and to dismantle her for not perfectly embodying feminism's ideals.

Like Ally, *Sex and the City*'s Carrie was also called ditzy and slutty for enjoying singledom and sex. The show premiered the same month and year as Ally's *Time* cover. It featured Carrie and her three friends dating, shopping, and living glamorous, single lives in New York City—lives that were relatively newly available to young women with careers and no spouses. As with *Ally McBeal*, the uniqueness of the show and its complex, independent women made it controversial. The characters were called vapid, trampy, and selfish. Carrie and company called pretty women younger than them bitches, foreshadowing possible catfights to come. Much attention was lavished on their wardrobes and how, in reality, they wouldn't have been able to afford them. "The show's shoe fetish has given way to more Fendi bags per episode than you can shake a trust fund at," ribbed the *Los Angeles Times*.

The women are criticized for their "hostility" toward the men they date, and for trying to emulate a prototypically male lack of attachment to casual sex. "For all the frantic coupling, no one seems to be having any fun," wrote conservative author Wendy Shalit. The sexual liberation the women of *Sex and the City* represented was duly shamed by the *Washington Times*: "The four protagonists, for all their cool urbanity, experience feelings of loss and sadness and loneliness that are real and typical for women in the age of liberation." The subtext seemed to be that they should pay for being free by being miserable.

Carrie and Ally were authorized to tell their own stories on mainstream television with the potent tool of voice-over. This was new power, but commentators ridiculed what women thought and talked about. Critics accused women's voice-overs of focusing on topics that were too boldly sexual, girly, selfish, and shallow. Ally struggles with work and loneliness, but her internal monologues and character are maligned as dopey, selfish, and too smitten with men to be taken seriously. *Sex and the City*'s Carrie saw much of the same criticism—a peek into her inner thoughts revealed a petty, boy-crazy, shopaholic narcissist.

Indeed, from television to boardroom to courtroom, women working in the 90s were subject to a galaxy of ridicule, and ultimately bitchification, stemming from the double bind. Display competence and you were criticized for lacking femininity. Appear too feminine and you were obviously incompetent. And often, rather than address these biases and the discrimination they allowed, the media narrative dug into stereotypes like the erotomaniac, the pushy babe, the Adorable Dope, and—of course—the bitch. It was as if these stock characters were the only kinds of women 90s America knew.

5

BAD MOM

The double bind was particularly acute for mothers, who faced an additional set of challenges and inequalities at work in the 90s. Policies and culture discouraged women who had earned powerful positions at the top of their fields from identifying as mothers in the workplace. If women were successful at work, it was assumed that they were neglecting their kids. Poor mothers on government assistance were accused of bilking the system to avoid work entirely, often because they were young, single, and black. These disparities and misconceptions coalesced to form entrenched stereotypes that dogged working women throughout the 90s. Bitchification expanded to include the "bad mother." Working moms, the culture concluded, couldn't be good moms. Women could publicly use their wombs or their brains, but not both.

Marcia Clark was one such mom. She filed for divorce from her husband, Gordon Clark, three days before Nicole Brown and Ronald Goldman were killed. Sure enough, during the trial,

Clark's home life interfered with her job in a way that became very public. It began when Clark asked Ito to reschedule the questioning of a key witness because she had to pick up her kids.

"I have informed the court that I cannot be present tonight because I do have to take care of my children, and I don't have anyone who can do that for me. And I do not want proceedings to go before a jury when I can't be here," Clark said. News reports emphasized her teary eyes and strained voice.

The defense was quick to hector Clark. She wasn't prepared to question Simpson's neighbor's housekeeper, Rosa Lopez, and was using her kids as an excuse, Cochran said. Surely the judge would see this request for what it was: a ploy for more prep time. Some mother Clark was, scapegoating her kids.

Clark's estranged husband publicly rebuked her mothering skills. He was suing for custody and tried to win sympathy for his case by claiming Marcia "misled" the court. "She has no childcare problem . . . I think it's inappropriate to use our children as an excuse in court," he said.

"While I commend her brilliance, her legal ability and her tremendous competence as an attorney, I do not want our children to continue to suffer because she is never home, and never has any time to spend with them," Gordon Clark said in a statement. "I have personal knowledge that on most nights she does not arrive home until 10 p.m., and even when she is home, she is working." The statement echoed Donald Trump's assertion, made less than a year earlier, that "putting a wife to work is a very dangerous thing." He blamed his wife Ivana's job for the dissolution of their marriage.

Not only were Clark's opponents in court and the presiding judge questioning her fitness as a parent, her ex-husband took these complaints—that she was a workaholic mother who neglected her kids—to the press. Gordon Clark seemed to invite the public to comment on his custody battle, and they did. The tension between women's work and domestic responsibilities reentered the national conversation.

"Sorry, Marcia, Kids Come First," scolded the *New York Daily News*. "What's the trouble if Clark loses custody for now?" asked the *Chicago Sun-Times*, as if losing custody of her children was akin to losing the house keys. A writer harrumphed that Clark wouldn't be "winning any Good Mother awards this year" in the *Boston Herald*, and speculated that her work prevented her from reading to her kids. "'Nurturing' is not a word in her lexicon," said another paper. "People were weighing in on whether or not I should get custody," Clark says. "It was like, are you kidding me?" Like her makeup, legs, and sexuality, Clark's motherhood would become a national fixation during the trial.

FATHERS' RIGHTS

Clark joined the long list of mothers savaged for pursuing careers. Her public battle dovetailed with a handful of newsmaking custody cases in the 90s in which mothers lost their children as a perceived punishment for their success. A lawyer in the office of Senate Judiciary Committee chairman Orrin Hatch publicly lost custody of her kids to her husband, who earned less than she did. The judge reasoned that her career seemed to take precedence over her kids, so therefore they were better off with their father. In another much-publicized case, a Michigan woman lost custody of her child to the birth father, with whom the baby had never lived, because she had enrolled in college and intended to put the child in day care while she was in class. In a similar case the year prior, a child was taken away from a mother whose boyfriend watched him part-time while she attended law school.

Women's groups expressed outrage that these women, and countless others, were so severely penalized for working hard to create better circumstances for their families. A 1994 study by the Families and Work Institute found that relatives didn't actually care for children any better than paid babysitters; in fact, they were sometimes worse. In the wake of these cases and this study, women's advocates and custody experts criticized judges for holding "lingering bias that penalizes mothers in custody disputes for working or going to school."

The fathers' rights movement shot back that custody rulings had too long prioritized mothers. "Any and all excuses are used by the justice system to perpetuate the myth that women should have the children in a custody case at any cost," wrote a father in the *San Jose Mercury News*. Dads had become disposable and scores of moms now practiced "dump a dad, get a check," according to a column in the *Washington Post*.

Fatherhood advocates seized on the plight of nanny iconoclast *Mrs. Doubtfire* to represent their cause. The 1993 film stars a divorced father, played by Robin Williams, who disguises himself as an elderly British nanny to surreptitiously spend more time with his kids. The story demonizes the career-oriented mother, played by Sally Field, who seems to exist solely as a blockade between Williams and his kids. This was precisely what Clark was doing, the dads' rights set charged. She was just another Sally Field, obstructing a good dad from seeing his kids. States like California and Maine proposed "Mrs. Doubtfire" laws, which prioritized fathers over other babysitters when mothers needed childcare, no matter the circumstances.

Bias against moms who worked outside of the home was deeply ingrained in the justice system. A 1990 survey revealed that half of Massachusetts probate judges believed mothers should be home waiting for their kids after school, and that pre-schoolers were somehow at risk if their mothers worked. This was the cultural climate complicating the lives of working mothers in the 90s, and Clark's celebrity put her mothering in the spotlight.

While it's hard to argue with divorced fathers wanting larger roles in their children's lives, that wasn't always the driving motive. Legal experts at the time called many of these cases something else. They said that fathers who didn't want to pay child support would often sue for custody to avoid payment. Or they sued when they stood to profit—for instance, when their wives earned more than they did, as was the case with Marcia and Gordon Clark.

Some commentators and Clark herself saw her ex-husband's custody case as a grab for the cash she was sure to earn post-trial. Gordon Clark accused Marcia of spending frivolously, on new clothes, shoes, and hairstyles. "She needed more money to improve her Hollywood glitzy image," he said. In an exclusive interview he gave to *Newsweek*, Gordon painted Marcia as not only a negligent mother, but an unfeeling miser. "She's saying, 'Not only can you not be with your kids, but I'm going to hire babysitters and you have to pay for them,'" Gordon said. "It made me feel like she wanted me to be a bank and not a father."

Even now, Clark bristles at all of this. "There's no such thing as a pleasant custody battle, but it was a hideous thing to be going through at the time. To be attacked because I'm busy in trial, which is a finite thing. It's going to end. I'm home to put them to bed. I'm home for dinner. And then I go back to work at the house." Plenty of women related to this dynamic in the 90s, as they would now.

NANNYGATE

As Marcia Clark learned, a mother with a powerful job is an easy target. But it wasn't just professional women's achievements that rankled the masses in the 90s; it was also that they were seen as sauntering off to successful, glamorous careers while paying other women to watch their kids, thereby dodging their primary responsibility of motherhood. Public anger at professional women who left their kids with caregivers was undoubtedly fueled by class resentment at the six-figure lawyers and politicians who could pay for private childcare while their low-wage and hourly counterparts could not. Condemning successful women for their childcare decisions was tantamount to saying they didn't belong in the workforce at all. Rather than take aim at the country's childcare crisis, the societal narrative pitted mothers against each other.

Zoë Baird would have been the first woman attorney general, and the reason why she wasn't is a well-known cautionary tale. The Connecticut attorney probably never thought that her career would be derailed by a childcare decision, or that her name would be associated with words like "failure" and "setback." Baird had long been a corporate lawyer, holding big jobs at General Electric and Aetna before President Bill Clinton nominated her to his cabinet.

The announcement took place on Christmas Eve 1992. A couple of weeks later, the story was leaked to the press that Baird and her husband had employed undocumented immigrants—"illegal aliens," as they were commonly called at the time—in their home. Baird had revealed this detail to the Clinton transition team when they vetted her and was told it was no biggie. She had paid a fine for the violation and back taxes for the employees before her nomination. But vocal resentment still brewed.

Critics speedily attacked Baird for being rich and aloof—a robo-yuppie, according to *Rolling Stone*. Numerous reports printed her salary as an Aetna attorney to prove her out-of-touch social status; her earnings topped $500,000 per year, eclipsing the salary of her husband—a Yale Law School professor—fivefold. Their New Haven house was said to look "a little like Monticello," and critics sneered at what they called her employment of "servants" to maintain a household of three. Baird's law career, salary, and childcare choices were presented as failed feminism. Rather than laud the accomplishment of female attorneys like Baird, her detractors claimed that the movement had promoted "the equal right of women to become yuppie power-lawyers alongside men," observed Robert Kuttner in the *Washington Post*.

In the early 1990s, it was commonly thought that hiring childcare was akin to outsourcing motherhood, and people balked that someone as wealthy as Baird would do it on the cheap. This revelation sparked a faux mommy war. Pitting different kinds of mothers against one another was the only way for many to understand the Baird conundrum. The cultural debate about motherhood did not allow for the role to be hard for both rich mothers and poor ones.

"Widespread anger" at Baird was reported among working

mothers who earned less than her, and among mothers without careers who worked at home caring for kids. Baird was a pariah because she was perceived to have the resources to make child-care easy for her family (never mind that, as any working mother will tell you, leaving your infant with a stranger is anything but). Unfortunately, Baird's story did not spur a needed discussion about the real villains of Nannygate: broken immigration laws and the country's childcare crisis, which was intensifying as more women joined the workforce and tacked on additional office hours. Instead, activists, politicians, and the media faulted Baird's success and blamed her for the difficult decision that made it possible for her to have a thriving career in the first place.

Baird told her side of the story and apologized at a hearing before the Senate Judiciary Committee in January 1993. She was forced to defend the decisions she had made for her child, and her motherhood, while a panel of mostly old, white men scolded her. Inarguably, Baird violated the law by not paying taxes and hiring illegal immigrants. But the emotional intensity of the discussion suggested that much more was at stake, that this controversy was really about what makes a good mother.

Baird revealed that she had been searching for childcare for almost two months so that she could return to work, and admitted she had acted more like a mother than an attorney general. To which committee chair Joe Biden replied, "It is amazing, the ability of the human condition to rationalize, to justify what you know is not right."

"There are . . . millions of Americans out there who have trouble taking care of their children . . . with one-fiftieth the income that you and your husband have—and they do not violate the law," he chided. His colleague, Wyoming Republican Alan

Simpson, highlighted that women felt they lacked Baird's life "advantage." As Baird sat opposite the committee members, one couldn't help but recall Anita Hill being interrogated by many of the same senators in the same grand hall a little more than a year earlier. And several media accounts did note the similarities between Baird's and Hill's congressional testimonies at the time. The committee treaded far more lightly with Baird after being lashed for how they treated Hill. But together, Baird and Hill symbolized the threats that working women in the 90s faced: sexual harassment and discrimination, even at the highest career levels, and the many challenges and attendant judgments that come with being a working woman. In both cases, instead of blaming insufficient laws, wrongheaded politicians, and broken systems, individual women were held responsible for national problems.

Women's groups were outraged that Baird was the only cabinet candidate queried about her babysitter, while none of the men had been. They claimed she was being held to a double standard: Baird's life and choices were being inspected with utmost scrutiny, far more than those of men who had already been confirmed. NOW president Patricia Ireland hit the White House hard on the double-standard point, reminding them that Secretary of State Warren Christopher claimed he didn't know the military had spied on antiwar protestors during Vietnam, despite government documents that proved he did, and yet he was speedily confirmed. Baird's childcare arrangement was a trifle compared to far more serious accusations. Some in the media agreed. "The reaction to Baird's admission, including calls for her to withdraw . . . far exceeded the response to inconsistencies in the records of other nominees," explained the *Boston Globe*.

NOW's six hundred affiliates launched a campaign to demand that confirmed cabinet members volunteer their own domestic employment records, since they hadn't been asked to do so by the administration. NOW clarified that not doing so would mean Baird was being held to a different standard as a woman appointee, and that not appointing a woman to attorney general would render Clinton as unprogressive as his Republican predecessors. Surprisingly, one cabinet official admitted he'd unknowingly done just what Baird had. Commerce Secretary and longtime Clinton confidant Ronald Brown revealed on *Meet the Press* that he had neglected to pay his house cleaner's Social Security tax. The White House stayed silent and "declined to even rebuke a male cabinet member," *Newsday* reported. Already confirmed, Brown kept his job. When Labor Secretary Robert Reich was asked about his own domestic employees, he sniffed, "I'm not here to talk about babysitters."

Public pressure mounted for Baird to withdraw. Clinton apologized and said he had rushed things. But this was not the case for the next proposed appointee, New York judge Kimba Wood, who was asked by Clinton himself during her vetting if she "had a Zoë Baird problem." After reviewing her financial records, the White House found that she had also employed an undocumented caregiver. Wood maintained that she had followed the law and paid all the proper taxes. It had all been aboveboard, but the administration feared recriminations after the Baird debacle. It was the whiff of a nanny problem that put the kibosh on Wood's prospects. Aides leaked that Wood had trained as a Playboy Bunny while studying at the London School of Economics, even though she'd quit after five days and didn't earn a paycheck. Nobody minded that part of her record, *Newsday* assured readers. And there wasn't a backup contender immediately available. By compelling Baird and Wood (who was never nominated) to withdraw from consideration, Clinton sent the message that mothers were not welcome in top White House jobs. Critics charged sexism. Many blamed Hillary Clinton, stating that her posturing to get a woman in the job had caused the mess in the first place. "There's no doubt that Wood—and Baird—were propelled forward by Hillary's network," *Newsweek* concluded.

WORKING-CLASS WOMEN

What fans loved about the sitcom *Roseanne* was that it drove an awl through the domestic-goddess trope. The show centered on the Conner family and its matriarch, who spit and cursed while she cooked and cleaned. Roseanne was both mom and housewife, while mocking the roles and their niggling humiliations. The concept for the show was based on the stand-up routine that comedian Roseanne Barr had performed on the road for eight years, featuring a self-proclaimed "fierce working-class domestic goddess" who was part lioness, part nudnik. Writing in *New York* magazine, Barr called the show "television's first feminist and working-class-family sitcom (and also its last)." She was such an oddity and curiosity that, in 1989, she broke a record for appearing on the most magazine covers.

The sitcom pointedly slammed the degradations of motherhood while capturing many of the role's essential truths. Viewers were relieved to see a more ribald, honest depiction of home life than typically shown on television. This was especially true a year after the debut of *Full House*, which sympathizes with a parent caught between work and the minutiae of raising a family, except that, in a departure from the norm, the parent at the center of the action in *Full House* wasn't a woman but a man. Iconic sitcom mothers would never have gotten away with Roseanne's caustic one-liners like, "If the kids are still alive when my husband gets home, I've done my job." Asked by one of her children why she's so nasty, she replies, "Because I hate kids and I'm not your real mom." When her daughter offers to jump off a cliff, Roseanne asks, "Why don't you take your brother and sister?" On the show, her girls were confronted with the troubles plaguing actual 90s girls like low self-esteem and teen pregnancy.

Her wife aptitude was also questionable. "I think women should be more violent, kill more of their husbands," Barr told the *New Yorker*. In the 90s, the idea of childhood innocence was still treated like fragile crystal, and many sanctimonious critics expressed disgust with Barr's jokes. They called her an "evil TV mom" and worried that the show was contributing to the dissolution of family values. "Her raucous antics . . . are calculated affronts to middle-class propriety," observed the *New York Times*. The *National Review* described her as "a cunning marketeer who has figured out how to parlay a form of vulgar reverse sexism into stardom." But plenty of viewers, parents, and families related to Roseanne's dark humor and sarcasm. These tools were also potent devices to a generation of 90s kids ever skeptical of trying too hard and being told what to do.

While white mothers who worked—like Roseanne, Clark, and Baird—were pilloried for appearing to shirk domestic and motherhood duties, unemployed or underemployed black mothers were shamed for failing to work and staying home with their kids. In the 90s, politicians and the press furthered a derogation of black women: the "welfare queen," who they alleged birthed scads of babies, scooped up government checks, and skipped work to chug champagne in her Cadillac. The archetype of the welfare queen was inspired by the racist notion that black women were lascivious, uneducated, and lazy. This disparagement, and the image it created of undeserving mothers riding a government gravy train, inspired laws that censured women's home lives.

Ronald Reagan demonized the welfare queen in the 70s, and by the 90s, she had become what social psychologists dub a "narrative script," a story told not only to explain ingrained behavior but also to predict it. "She is portrayed as being content to sit around and collect welfare, shunning work and passing on her bad values to her offspring," wrote Patricia Hill Collins, author of *Black Feminist Thought.* "The welfare mother represents a woman of low morals and uncontrolled sexuality." This stereotype pinned to poor black women convinced many that these women would connive and manipulate the system to get federal assistance they did not deserve.

The news media in the 90s mostly depicted African Americans unsympathetically, according to Yale political scientist Martin Gilens, who studied welfare stories in print and television news from the 1960s to the early 1990s. Gilens found that 62 percent of stories about poverty in major newsmagazines featured African Americans, and 65 percent of television reports about welfare did the same. Such portrayals stoked America's belief that the bulk of welfare recipients were black women, even though they only comprised about 10 percent of them in 1998. The majority of welfare recipients were actually children. This racist assumption echoed the Jezebel and Sapphire tropes used to disparage Anita Hill and *Living Single*'s powerful working women. Sapphires and Jezebels connived and manipulated in the office and bedroom respectively, while the welfare queen did so from her living room sofa.

The welfare queen was a potent motif that fueled the media narrative around supporting the welfare overhaul of 1996. The law itself promised to "reduce non marital births and encourage marriage," with financial incentives to states that accomplished these goals without raising abortion rates. Tying government assistance to a moral framework is part of the law's legacy. And that didacticism was predicated on reducing black women to idle childbearers.

FAMILY VALUES

The fictional Murphy Brown's choice to become an unwed mother sparked a national debate. Many fans wanted Murphy to remain childless and to terminate her unplanned pregnancy to affirm that motherhood wasn't all women's holy grail. Murphy's decision to become a mother kicked off "one of the most bitter chapters in the 1990s culture wars." She gave birth to her son in the finale of the 1991–1992 season, which drew thirty-eight million viewers—more than baseball's World Series final that year.

Her choice became a referendum for feminists. Plenty were dismayed to see Murphy defect to motherhood after rejecting it, and accusations that Murphy had failed women mounted. In a moment thick with meaning, Murphy sings softly, "You make me feel like a natural woman," to her newborn, tears springing from her eyes, "implying that she had been unnatural before," wrote author Susan Douglas. "Natural Woman" was Murphy Brown's theme song, according to the show's creator, Diane

English, mocking the idea that motherhood could make a woman "natural" and softening Murphy's hard image all at once.

In May 1992, Vice President Dan Quayle made a campaign stop in California at the Commonwealth Club of San Francisco. There, in a moment of moralizing about the country's dearth of family values, he blamed the fictional Murphy Brown and chided her for having a child without a husband.

"Bearing babies irresponsibly is wrong," he began. "It doesn't help matters when primetime TV has Murphy Brown, a character who supposedly epitomizes today's intelligent, highly paid professional woman, mocking the importance of fathers by bearing a child alone and calling it just another lifestyle choice . . . Some things are good and other things are wrong." And with that, Murphy went from "lovable bitch" to aberration.

These words from the vice president—who hadn't even seen the show, it turned out—set America aflame. His supporters and traditionalist family groups savaged Murphy and the single motherhood she glorified. "Out-of-wedlock birth"—the term itself pejorative—was "deviant behavior" that was wretchedly being "normalized" by the culture, moralists argued. They were also talking in racist code about the illusive welfare queens, of course. The show was celebrating the disintegration of the American family and was contributing to a national campaign to accommodate deviant behavior.

As Quayle noted with indignation, Murphy Brown was also blunting the role of fathers and ignoring studies and social indicators that favored two-parent households. The Catholic League argued for a "restoration of family values" that Murphy had trampled with her decision. Fatherlessness was "devastating . . . as even

Murphy Brown would admit if she were as good a reporter as her TV show depicts her," wrote one detractor in the *Chicago Tribune*. The country would hear echoes of this logic in the fathers' rights movement that, soon after, would fault Marcia Clark for trying to maintain custody of her children while going to work.

Since Quayle hadn't watched the show, how did he even know enough about Murphy Brown's domestic choices to criticize them? It turned out his speechwriter, Lisa Schiffren, was to blame. Even though a powerful man had said the words, she had written them and rendered Murphy "a weapon in the right's attack on single motherhood." She had read about the show and mentioned its willful celebration of single motherhood to her boss. He had liked the idea of critiquing Brown in his speech. After Schiffren left her post as Quayle's speechwriter, Candice Bergen wrote a letter to the editor of the *New York Times* criticizing Schiffren for being a self-proclaimed "full-time mother of two and an occasional writer. Not every woman has the luxury to make that choice," Bergen jabbed. Murphy's single motherhood might have been morally heroic and worth defending, but since a woman had spearheaded the national backlash, Schiffren's choices were fair game to skewer, too. This was characteristic of the tenor of the debate between mothers who worked in offices and mothers who worked at home in the 90s—both roles were seen as unyielding, permanent identities. This lingers today. Each seems to criticize the other's choices in order to reinforce their own.

Plenty of *Murphy Brown* fans laughed at Quayle's tone deafness. The show's writers and producers used the sitcom to respond to the vice president directly. In one episode, a disheveled Murphy watches the vice president lambaste her on television while home with her newborn. She later addresses Quayle from her anchor desk. "Perhaps it's time for the vice president to . . . recognize that, whether by choice or by circumstance, families come in all shapes and sizes. And ultimately what really defines a family is commitment, caring, and love."

Quayle tried to walk back his words, emphasizing that he never meant to criticize single mothers and that his sister and grandmother were single mothers, too, because they were divorced. The real vice president even sent a stuffed elephant to Murphy's fictional son. The show's crew said it would donate the toy to a homeless shelter. Quayle's press secretary later acknowledged to the *Los Angeles Times* that his boss was battling a fictional character, and how that might seem nuts. The flap became too politically hot for Quayle to handle. He eventually dropped "family values" from his stump speeches.

Murphy Brown had a unique dais from which to comment on politics. It riffed on the Anita Hill hearings in the wake of that controversy, for instance. Doing so made the show a must-see for the Beltway elite, who emptied the capital Monday nights "to head home and watch and see if they were name-checked," English told me. Her Los Angeles office features framed front pages of newspapers covering the *Murphy Brown*–Dan Quayle spar. What was truly significant, according to English, was how the debate around Murphy's baby impacted that year's presidential election, which would end twelve years of Republican control of the executive branch. This was what English had hoped to do all along.

"My goal from the beginning was to erase the line between fiction and nonfiction," she says. "When the vice president is talking about Murphy Brown as if she's an actual person, that was the definition of success for me. The presidential election got turned a bit on that debate." Antiestablishment candidate Bill Clinton began to look more like the future, while George Bush and Dan Quayle seemed increasingly fossilized in their hostility toward women, working mothers, and the nation's cultural evolution.

By 2003, a two-year survey of more than three thousand women found that equal pay and childcare were still two top concerns for women that had gone unaddressed by the government and employers. Domestic violence was number one. It took the 90s ending for society to realize that "one Carly Fiorina doesn't change the world," as gender politics researcher Shere Hite, author of the book *Sex & Business*, told the *New York Times* in 2000. Hite was referring to the Hewlett-Packard CEO, who was among the first women to run a Fortune 100 company. One Marcia Clark and one Murphy Brown didn't change the world either. Women were so beat down by work culture at the close of the decade that they were opting out of the workforce as they had kids, or starting their own companies to avoid it, because they "feel it's better to put their energy into that than into fighting the old system and being derailed into the sidelines," Hite said.

White or black, working or not, women with children quickly found that it was impossible to be a good mother in public in the 1990s. But while the gender struggles of women in the workforce were daunting, they paled next to those of the women who tried to breach the corridors of power in Washington, DC.

6

FIRST BITCH

In the 90s, three women politicians created truly national profiles by reaching for power and insulting the patriarchy along the way. The nation cheered and lampooned them for their drive, their ideology, and their hair. They were infamous for being different, and famous for being first. Hillary Clinton, Madeleine Albright, and Janet Reno served in different capacities in the Clinton administration, which defined American politics throughout the decade, and perhaps ever since. Reno was the nation's first woman attorney general, and Albright the first woman secretary of state, while Hillary Clinton was the country's First Bitch.

In 1991, the Anita Hill hearings revealed how absent from and victimized by political power women really were. The event inspired women to seek political office on both the local and national levels. That explains why 1992—when four women won US senate seats, raising their ranks from 2 percent of the chamber to 6 percent—became known as the "Year of the Woman."

The Hill hearings had galvanized women politically. Still, the sitting president rooted against them. "This is supposed to be the year of the women in the Senate," President George H. W. Bush said at a presidential debate in the lead-up to the election. "Let's see how they do. I hope a lot of them lose."

Dianne Feinstein and Barbara Boxer won seats in California, Patty Murray in Washington, and Carol Moseley Braun in Illinois. Ask anyone what the "Year of the Woman" means, and they probably have no idea. Even the women elected to the Senate that year weren't thrilled by the branding. "Calling 1992 the 'year of the woman' makes it sound like the 'year of the caribou' or 'year of the asparagus,'" said Senator Barbara Mikulski of Maryland. "We're not a fad, a fancy, or a year."

While these senators blazed a path to power for women in Congress, they didn't attract the kind of controversy and national fame that Clinton, Albright, and Reno did. This trio negotiated with foreign powers, shaped policy, and managed large staffs. Their powers, combined with their sheer visibility in the new realm of 24/7 news coverage, made them more threatening than the women senators. Clinton, Albright, and Reno were powerful not by membership in a legislative body, but in and of themselves. As such, their authority was endlessly questioned and feared. Media narratives and even these women's colleagues perpetuated the notion that power in their hands was dangerous.

LADY MACBETH

As Bill Clinton assumed the presidency in 1992, he promised American voters their first chance at a woman president. What he meant was, the American people were in for a bonus brain, a presidential-quality civil servant in his wife, Hillary. Supporters called it "two for the price of one."

During the election, the accusation that Bill Clinton had engaged in a twelve-year extramarital affair with singer Gennifer Flowers marked a new kind of White House bid: one tinged by tabloid prurience. This revelation was said to steer Bill Clinton's campaign to "the brink of implosion," but it also unleashed waves of criticism on his spouse. Nearly every shred of evidence pointed to an affair. What wife stays with a husband who cheats on her for more than a decade? This was Hillary Clinton's introduction to the nation.

The couple plays offense during a joint *60 Minutes* interview, which aired after the Super Bowl in January of 1992. They are seated on a couch with hands intertwined. The slithery Bill refuses to admit guilt or innocence. He pleads with the American people to consider their own private, imperfect marriages, while scolding the licentious press for clawing their way into his bedroom. By the end of it, you're with him, believing *60 Minutes* interviewer Steve Kroft is the rake. This notion is furthered when Hillary Clinton said later in an interview that Kroft asked questions kindly when the couple was in studio, but that cameras reshot the host asking in a tougher manner once they had left.

Kroft, coming up empty, makes a final play for dirt. "You've seemed to reach some sort of understanding and arrangement," he says of their marriage, insinuating that political aspirations and backroom deals soldered their bond. In typical Clintonian aw-shucks, Bill says, "Wait a minute!" three times and then, "You're looking at two people who love each other. This is not an arrangement or an understanding. This is a marriage." Hillary

Clinton follows, in a voice inflected with Southern twinges since vanished, "You know, I'm not sitting here—some little woman standing by my man like Tammy Wynette." She was referring to the country music star whose hit song, "Stand by Your Man," celebrated deferential women. Hillary extends her drawl on these words, further jabbing the tear-streaked woman done wrong in the country ballad. Pundits would later call it her condescending "black-cent" that she used to hoodwink Southern and black audiences into thinking she was more bumpkin than boss.

"I'm sitting here because I love him, and I respect him, and I honor what he's been through and what we've been through together," she says. "And you know, if that's not enough for people, then heck, don't vote for him." He rubs her hand vigorously, as if washing it with soap. It was the start of what some journalists called the "Lady Macbeth framing" of Hillary Clinton. She had been masterminding his campaign and speaking to the press as if she were the candidate herself. Now she was perhaps sweeping an affair beneath the carpet. Hillary was the maniacal puppeteer manipulating her husband so she could ascend the throne, like the Bard's famous character. "I could tell she was being seen as a bitch," said MTV News reporter Tabitha Soren, who was the face of the network's 1992 election coverage, dubbed "Choose or Lose." "But it didn't bother me, because I thought she'd be seen as intelligent as she was bitchy."

The Clintons were not in Arkansas anymore. The Tammy Wynette dig was broadcast nationally, piped into the homes of millions. Many viewers were dubious: Wasn't Hillary Clinton doing exactly what she had supposedly rejected, standing (well, sitting) by her philandering man in a television studio, wearing a modest headband and turtleneck up to her chin? British press had pinned her "the meek, mild, wronged wife." She was cast as an elitist for her achievements—*Yale Law Journal* editor, counsel for Nixon's impeachment, corporate attorney. Clinton *was* standing by her man, and for her own selfish gain so she could become copresident without being elected. Writer Christopher Hitchens explained that Hillary "knowingly lied about her husband's uncontainable sex life and put him eternally in her debt."

The charge that she was riding on his coattails would reemerge in the wake of the Lewinsky scandal, when the Clinton marriage was called "the longest, slowest, most painful car crash in marital history," and Hillary was blamed both for inciting the affair with her career and ambition and for not leaving her husband. With the Wynette comment, Hillary had spit on the honest, small-town values America held dear. She was mocking those with less education and opportunity, and "appearing to show contempt for women who work at home," journalist William Safire wrote.

Conservative women leapt to criticize Hillary's inflammatory remarks. Outgoing First Lady Barbara Bush, who perfected the "everybody's grandmother" image, declared that she would gladly be introduced to crowds with "Stand by Your Man" blaring in the background. Tammy Wynette, the First Lady of country music herself, was "mad as hell" and demanded an apology. "I can assure you, in spite of your education, you will find me to be just as bright as yourself," she told Clinton through a reporter.

Two months later on the campaign trail, Clinton stepped in it again. "I suppose I could have stayed home and baked cookies and had teas, but what I decided to do was fulfill my profession, which I entered before my husband was in public life," she told reporters asking about her law firm job. She went on to say that feminism granted women the choice of work, at the office, at home, or both. But that wasn't the sound bite. Critics reproved Clinton for mocking housewives and stay-at-home mothers. Republicans cast her as a "wild-eyed feminist who equated marriage with slavery," and a danger to American families. Her views smacked of elitism. She had attacked women who forgive men their grievances, then stay-at-home moms, all in a matter of weeks. These comments—which seemed to flaunt her education, achievements, and career success, and denigrate domestic women as lazy dependents—angered housewives and scared their husbands. Fear of her was commercialized with the release of Hillary Clinton nutcracker dolls, legs spread wide to crush walnuts—or cojones. (They are still for sale on Amazon.)

Suddenly, the topic Bill Clinton was asked about most on television and in interviews was not the economy or foreign policy, but his wife. She introduced him at rallies, but seemed unwilling to relinquish the podium. "Perhaps never in a presidential campaign has the candidate's wife become such a strong symbol of the campaign's strength and weakness," journalist Ted Koppel said.

The promise of a "twofer" presidency wasn't a partnership at all, but a nefarious trick to launch Hillary into the Oval. Detractors began to search Hillary's appearance for signs of menace. You could "detect the calculation in the f-stop click of Hillary's eyes," *Vanity Fair* surmised. She took on a Stepford sheen: "Lips pulled back over her slightly jutting teeth, the public smile is practiced; the small frown establishes an air of superiority; her hair looks lifelessly doll-like." Most long-serving politicians commit gaffes, and the gaffes are eventually forgiven; just ask Joe Biden. And yet, somehow, more than twenty-five years after the Wynette and cookie incidents, the deep hatred of Hillary Clinton suggests that she is still paying for her words.

"Too intense, explosive" Hillary needed to be contained to salvage her husband's campaign. Advisers began with her wardrobe. The campaign had already dressed her in girlish headbands, like the one she wore in the *60 Minutes* interview. Post cookie comment, her closet filled with softening pastels, like Marcia Clark's two years later during the O. J. Simpson trial. This tactic was no match for the entrenched caricature of Hillary that had already formed in the minds of Americans. A *New Yorker* cartoon satirized how women reacted to Hillary's clothes—a shopper trying on a jacket tells the sales clerk, "Nothing too Hillary." The wardrobe makeover attempted and failed to whip a "loving wife and mother" out of the "ambitious yuppie from hell," as ABC News put it.

Perhaps entering the kitchen she had scorned would soften her image? In July of 1992, she committed to "bumping spatulas" with First Lady Barbara Bush in a cookie bake-off for *Family Circle* magazine. The *New York Times* reported that Clinton was competitive, even with sweets, telling a group of congressional wives that "she was going to all-out win." And indeed, her chocolate chip cookies, which she said contained healthy oatmeal and oil, beat Barbara Bush's buttery shortbread in the era when margarine was queen.

"A WIFE WHO DOMINATES HER HUSBAND"

After Bill Clinton was elected, rumors swirled that Hillary would usurp undue power in the White House. She was the first First Lady to have a postgraduate degree and a career. It was reported that she would have her own West Wing office with more and higher-level staff than Vice President Al Gore. On the trail, reporters had asked if she wished to replace the senator from Tennessee as her husband's vice presidential nominee. "I'm not interested in attending a lot of funerals around the world," she joked. Then she grew serious. "I want maneuverability . . . I want to get deeply involved in solving problems."

America grew unsettled as the media accused Hillary Clinton of overstepping her bounds, speculating that she would hold a cabinet position, choose cabinet members, or pass laws. Indeed, Bill was whom the country had elected, and one could argue that the Clintons asked for criticism by promoting themselves as a "two-for-one" deal. But sexism was present, too, if often disguised as traditionalism.

"It's not that she's an accomplished modern woman," a Republican political consultant told the *New York Times*. "It's just that she's grating, abrasive and boastful. There's a certain familiar order of things, and the notion of a coequal couple in the White House is a little offensive to men and women." A 1992 *Vanity Fair* poll had found that 44 percent of respondents thought Hillary was "power-hungry," 36 percent said she was "too intense," and 28 percent agreed she was "a wife who dominates her husband."

The press demanded that Bill Clinton make clear to the American people how he intended to deploy his wife. "Voters should know beforehand what sort of First Ladyship is in store. Would she work on the outside or the inside?" William Safire wrote. The questions gave voice to the insecurities of the voting populace. America "wants a First Lady to be an adjunct to the man that they elected, but they have no control over her, and that, I think, causes a great deal of fear," said author Patricia O'Brien, speaking on ABC's *Nightline*.

To blunt these fears, Clinton marched her pacification tour of baked goods into the White House. She served reporters cookies after a Christmas Eve news conference announcing cabinet appointees, including Zoë Baird and Joycelyn Elders. News outlets called her move "uncharacteristic," but the Clintons surely knew that handing out the desserts that got her in trouble would grab attention. She continued her confection offensive to telegraph that she had been tamed and was no longer a threat. She couldn't resist cake! She confided to the *New York Times* that "rich, rich, rich chocolate cake with thick chocolate icing" was her vice. "Chelsea and I love chocolate," she said. As presidential historian Gil Troy put it, "She was, essentially, fired from her public role as Bill Clinton's sentry—and backbone."

The sweets quickly faded from memory as Hillary Clinton's influence in the White House seemed to grow. By 1994, Hillary was working on a healthcare bill with a private committee. Detractors accused her of wresting and then squandering power that didn't belong to her. What she presented to Congress was regarded as a flawed rookie effort. Her proposal was never put to a vote, and when it was deemed a flagrant failure, she was wholly blamed. Accusations included that she had been too know-it-all to incorporate doctors into her plan, that she wasn't qualified to craft high-level policy, and that she wasn't a strong enough coalition builder. The main charge, though few said it outright, was that people just didn't like or trust her. In 1995, Hillary Clinton set tempers to boil when she called out the sexism that had been working against her. "If I want to knock a story off the front page, I just change my hairstyle," she joked.

Healthcare reform was a meaty policy task that Bill Clinton had run on, and by leading the charge, "ultrafeminist" Hillary was accused of emasculating her "wimpy" husband. Her determination to rewrite law, to do the regular work of politicians, encouraged critics to joke that she was a man. A 1995 *Spy* magazine cover brought this notion to life. It featured a photo of Clinton with her skirt blown up to reveal men's underwear concealing a bulging penis. The cover line read: "Hillary's Big Secret." The masculine Hillary was "welcoming men to their role as the second sex," quipped the *Weekly Standard*.

POWER BEHIND THE THRONE

As First Lady, Hillary Clinton represented the "deeply American fear of the unaccountable power behind the throne," according to Troy, who has written about first families. "There are invisible trip wires around the president that First Ladies trigger," he told me, and while some First Ladies scoot around them, Hillary Clinton stomped them hard. She made headlines for appearing with her husband's trusted advisers, attending meetings with top brass, and voicing opinions on policy and government matters.

Troy believes that the more visible a First Lady is, the more likely she is to become a potential target. "There is an understood First Ladies' version of the Hippocratic oath: 'Do no harm.' They can have greater potential to do harm than good," he said. Hillary Clinton, who was incredibly visible during the campaign and transition into the White House, played into public discomfort with a First Lady's duty. Troy added that while reporters tend to long for modern, careerist First Ladies like Hillary Clinton and Michelle Obama, whose stars both rose long before their husbands', the country still much prefers the traditionalist model embodied by Melania Trump and Laura Bush. More cookies and readings, and fewer opinions and policy roles.

After Bill's reelection, Hillary morphed from a maniacal puppeteer into First Lady cliché. Hillary haters bathed in schadenfreude, as their target retreated to planning teas and the Easter egg roll. Gone, albeit temporarily, were the accusations that she was a "hall monitor" whose "offputting" "drive and earnestness" polarized the populace. She published *Dear Socks, Dear Buddy*, a collection of children's letters to the first dog and cat. She penned a weekly syndicated newspaper column on topics like mammograms, postpartum hospital stays, Big Bird, and trimming the White House Christmas tree, an eighteen-foot Fraser fir. One paper described the column as "nattering cheerily," having "sublimated her into a nightmarish amalgam of Lady Bountiful, Florence Nightingale and Betty Crocker."

Toward the end of the Clinton administration, Hillary managed to circle back to policy, helping to pass a federally funded children's health insurance program and a law to facilitate adoptions from the foster system. These were tremendous gains, but they were perceived as soft and unthreatening woman's work, which is likely why she was able to achieve them. Ruth Mandel, an expert on women's political history, put Hillary's dilemma this way: "When it comes to women, people are not ready to take more than a teaspoonful of change at a time."

MISTRESS VS. WIFE

When the Lewinsky scandal broke in January 1998, policy took a back seat once more. Pitting Monica against Hillary seemed a natural story line to the media, yet another catfight cage match that came to prominence in 90s news coverage. The prospect of a Mistress vs. Wife smackdown in the Oval Office was delectable. The media wondered who had the most sexual value in the eyes of the same man. In Barbara Walters's interview with Lewinsky, she asked several questions about Hillary Clinton that perpetuated the public desire for a catfight, including if Lewinsky felt in competition with Clinton; whether the president acknowledged Lewinsky when Hillary was around; and, egging her on, "Did you ever think about what Hillary Clinton would be feeling or might feel if she knew? Did you think about Hillary Clinton?" T-shirts that read "Lewinsky: I Get the Job Done When Hillary Can't" furthered Walters's suggestion that the two women were competing. Many, including Walters, seemed to shame the mistress and side with the wronged wife.

Plenty of people wanted to blame the wife, too. After all the perceived softening—a book of letters to pets, healthcare for families, homes for foster kids—criticism reemerged that Hillary Clinton had been deceiving Americans all along and was at her core too power-hungry to be a good wife. Clearly, she couldn't satisfy her own husband and was at fault for his adultery. His affair was payback for her education and ambition, and for being an ice queen.

Women in particular were quick to assume she stayed in her marriage for sinister reasons. She needed a man—this man—to further her career, they alleged. The affair proved she was a fraud. She was "unmasked as a counterfeit feminist after she let her man step all over her," wrote *Times* columnist Maureen Dowd. Her marriage was a farce, according to a *Vanity Fair* poll that found only 22 percent of people believed the Clintons had a "real marriage," while more than half called it "a professional arrangement." In *Salon*, journalist Jake Tapper termed Hillary Clinton "a shrew whose capacity for denial is equalized only by the pain she's suffered as a result of the fine print in her Faustian marriage contract." Bill Clinton allegedly rejected a drafted speech featuring an apology to Lewinsky and her family for their suffering in the spotlight, deferring to Hillary's "brass-knuckle guidance" and dismissing Lewinsky as "inappropriate" instead.

As the Clintons wrapped up their eight-year stint in the White House and speculation grew that Hillary might run for political office, her ambition continued to be rebuked. It turned out that liberal, successful professional women, just the type who should cheer Clinton and form her base, actually hated her, according to a report in the *New York Observer* titled "Meet the Smart New York Women Who Can't Stand Hillary Clinton." These women were developing a "grudge against Mrs. Clinton as a representative of their sex," the pink paper disclosed, because she was using a man to access the halls of power. These women just "couldn't relate to her on a personal level." Writer Fran Lebowitz called her "a very poor role model for girls" and "regressive" for marrying the president rather than becoming the president herself. She accused Hillary of pandering to the powerful and lacking ideas. Memories of her walking arm in arm with her husband to his impeachment were still fresh. Tammy Wynette, indeed.

"During my time in Washington, I heard Hillary Clinton called many things," wrote newspaperman Doug Thompson reflecting on the 90s in the online news site *Capitol Hill Blue*. "'Bitch' is one of the more polite terms."

JANET RENO'S DANCE PARTY

Will Ferrell pitched the skit "Janet Reno's Dance Party" in his second season on the *Saturday Night Live* cast. Female cast members believed that Ferrell had "a very female style of comedy," Yael Kohen, author of *We Killed: The Rise of Women in American Comedy*, told me, "meaning it was very character-based rather than joke-based." Eager to prove his comedy chops, he was after a "broad and physical role," and thought it would be funny to play a "large woman manhandling people," Ferrell told the *Washington Post*'s Liza Mundy in 1998.

Colleagues loved the idea of a dance party featuring a deep-voiced, awkward, tone-deaf, out-of-touch, purportedly lesbian attorney general who danced to the same song ad nauseam to gyrate away her mistakes. "Dance party makes Waco go away," Reno bellows in one scene, referencing the US government's attack on a religious cult's compound in Texas that left dozens dead. Ferrell described the Reno character he created as a "tough woman who lives in this make-believe world."

SNL in the 90s saw lots of men playing women, but few women playing men. "The sexism is in the fact that you have fewer roles for women if the men are also playing the female roles, and then the women are left playing girlfriends, mothers, or nuns—the traditional roles," Kohen said. "You have a limited amount of women you can play, and you don't get to play Janet Reno on top of it." Tina Fey calls out the *SNL* problem of casting dudes in drag in her memoir, *Bossypants*. "I remember thinking that was kind of bullshit," she wrote of an incident in which

Cheri Oteri was overlooked for a part in favor of Chris Kattan in a dress. "I think Cheri would have been funnier."

In the "Janet Reno's Dance Party" sketches, Ferrell achieved his recurring role. Eight dance parties aired in total from 1997 to 2000, helping launch him to stardom. Ferrell's Reno outfit resembles a Halloween costume. His wig approximates her short, flat hairstyle. He dons her face-filling glasses and trademark azure suit. Ferrell towers above the other dancers, like Reno did over her fellow cabinet members. He dances like a jerky carnival ride, throwing gawky elbows and knees.

Ferrell's Reno speaks in a voice that is evil-cyborg deep, dances awkwardly with teenagers, and barks orders at them, like "No mosh pit!" and "Shut UP! Shut it!" They are unfazed because Reno doesn't seem to have any real authority in these skits, only poutiness and fantasies. Describing his dramatic strategy for the character, Ferrell said, "I just sound the way she looks." Comedy about real people usually involves caricature—Alec Baldwin exaggerates Donald Trump's vulgar language impersonating him on *Saturday Night Live*; Kate McKinnon magnified Hillary Clinton's unbridled ambition. But the very fact that Reno is played by Ferrell calls her womanhood into question before the character moves or opens her mouth.

WACO

Ferrell's portrayal of Reno didn't materialize out of vapor. From the moment she arrived in Washington in 1993, she was a curiosity to be poked and gawked at. Reno was not the top choice for the cabinet post she would be the first woman to fill. She was the third, nominated only after the prior two succumbed to Nannygate.

Reno was the consummate outsider. She built her career as a county prosecutor in Florida, far away in both distance and culture from the nation's capital. She had managed to win elections for state attorney as a straight-shooting liberal Democrat in heavily conservative Republican Dade County, which impressed politicians across the country.

Jamie Gorelick, Reno's deputy attorney general, told me that Reno "led with her values, which meant that the people in the Department of Justice and outside it followed her just on the strength of her moral stance." She was called the most qualified of the president's cabinet officials and received a standing ovation at her swearing-in ceremony. Reno's approach to law enforcement was as foreign to Washington as tropical birds. A major city police chief with more than thirty years in the force called her "the most refreshing law-enforcement leader I've met in my whole career."

This patina of intriguing newcomer and America-defending cop quickly wore off. Reno had barely been sworn in a month when, in April of 1993, she ordered a tear gas attack to end the siege on cult leader David Koresh's compound in Waco, Texas. The leader of the Branch Davidian cult was said to be abusing and threatening the lives of women and children there. The strike accidentally started a fire that killed seventy-six people. Later, in a deposition, Reno called the Waco standoff—which began with a weapons raid that claimed the lives of four federal agents—"the most urgent issue I faced when I took office."

Forthrightly, Reno owned up to approving the plan, took responsibility for the death toll, and apologized sincerely to the American people. A bold admission of guilt and a heartfelt apology from a new government official stunned a populace accustomed to the evasion of wrongdoing that politicians perfect. Meanwhile, Bill Clinton was nowhere to be found, which gave the impression that he had leapt aside to watch Reno take the fall.

Reno called the blitz on Waco the worst day of her life. While onlookers commended her apology, their reverence didn't last, especially as more information about her decision was released. Her reputation morphed from earnest leader to murderous renegade. The label stuck—protesters at a documentary film about the events fifteen years later wore "Butcher Reno!" T-shirts.

SWAMP THING

Florida being Florida, Reno arrived in Washington with a colorful backstory. Having grown up in the Everglades, she was rumored to be intimate with reptiles, "rassling" alligators with her mama, and hiking through backwoods swamps. She strode into Washington shrouded in this mythology. The political establishment could tell that her look was gossip-worthy. To start, she was very tall—six foot one, to be exact, which elevated her over the heads of many men. She and the president look about the same height and wear similar haircuts at her swearing-in ceremony in March of 1993.

While women politicians usually stand out in the cacophony of dark suits by wearing bright colors, jewelry, makeup, and long hair, Reno did the opposite. Her wardrobe was filled with the considered pastels women lawyers wore in Florida. By avoiding jewelry and makeup, and wearing short hair and a staid palette, she actually drew more attention, "like a pileated woodpecker," according to the *Washington Post* article "Why Janet Reno Fasci-nates, Confounds and Even Terrifies America." The *New Yorker* scoffed at Reno's fashion sense with its "Selections from the Janet Reno Collection" cartoon published in October of 1993. It de-picts the attorney general in a series of panels, wearing cheap fabrics, an ugly mustard jacket, and a "mock-croc" belt. In one image, Reno sports a "scoop-front bra and bike short" with a "high-cut Danskin thong brief," posing with an exposed tummy like a thicker Jane Fonda. The joke was that Reno couldn't dress herself to save her life.

U.S. News & World Report published an illustration of Reno lassoing a gator. Another drawing represented her as a gun-toting Rambo. These cartoons exploited Reno's unfamiliar be-havior and lifestyle. She didn't have children or a partner. She was uninterested in parties and socializing, preferring to kayak or hike the Billy Goat Trail in nearby Maryland. "She was not terribly comfortable in the ways of Washington," Gorelick told me. "She acclimated to it, but it wasn't her natural environment."

Since Reno lacked the feminine qualities and life choices typ-ical for women, run-of-the-mill sexism wouldn't do. *Time* mag-azine speculated that Reno got the job in the first place because she had no husband, kids, or nanny, and thus "no Zoë Baird prob-lem." Instead, because she was awkward and tall, and had no dis-cernible romantic life, she was cast as a man in women's clothes. She had "a self-conscious hunch to her shoulders" and "awk-wardly dangling arms," according to a *New York Times Magazine* profile. Because she came off as tough after Waco, and because she openly challenged her boss, the president—like when she publicly pressed him to appoint her to a second term when it seemed clear that he would not—critics attacked her femininity.

For years, the late-night television fraternity spun jokes that Reno was a man. When she appointed an independent counsel to investigate Bill Clinton's presidential campaign finances—a ballsy move that endangered her boss—Jay Leno scoffed that it was her "toughest decision since boxers or briefs." Bill Maher compared her to former FBI head and notorious cross-dresser J. Edgar Hoover, whose dresses "fit her perfectly." David Letterman said he'd pose as Reno, but to do so he needed more shoulder pads. *Spy* magazine, known for sharp satire and comically doctored photos depicting powerful women as men, plopped Reno's head atop a gun-toting, Rambo-like body and titled the image "Mother Justice." To be sure, politicians of both genders are routinely mocked for their looks, but it wasn't just comedians and satirical publications that portrayed Reno as unfeminine. One magazine profile of Reno bore a section titled "Swamp Thing."

AUNT JANNY

The flip side of Janet Reno the man was Janet Reno the sissy. Her provincial nature, which at first made her interesting and unlike Washington's political cannibals, soon fed the theory that she was unqualified for her job. Rather than talk forcefully, the nation's top cop mumbled. Her voice was flecked with "slow Florida twang," rendering the second weekday "Tewsdee" and poetry "poitree," projecting the image of slow and stupid often assumed in the Southern drawl.

Reno's dedication to children and families—the product of years spent in law enforcement watching neglected and impoverished youth grow up to become criminals—was seen as weak. The nation's head law enforcer was likened to a social worker.

While Clinton's policies were filling jails, Reno advocated against mandatory minimums for drug offenses and championed the drug courts that she had successfully pioneered in Florida to reroute nonviolent offenders. It was a prescient policy, but she was painted as a crime apologist. Some suggested she was "more suited to be the Secretary of Health and Human Services," a demotion in power and influence.

Reno's simultaneous reputations for mannishness and weakness prompted speculation about her sexuality, which was rare for a United States attorney general, but not for a woman in power. Some believed her look to be "a sexual signal" that she was a lesbian. Jay Leno insinuated this when he took the liberty of superimposing her head atop Xena the Warrior Princess, the badass in armor with a gay following. *Time* quoted a friend of Reno's who said she "would love to have a relationship with a man and have children," but that it was tough to find someone "not threatened by a successful woman."

Senator John McCain joked in 1998 that Chelsea Clinton was so ugly "because Janet Reno is her father." He allegedly apologized to Bill Clinton for the smear a decade later, but not to any of the women.

Reno's gayness was so often assumed that she addressed it publicly, calling herself an "awkward old maid who has a very great attraction for men," not women. Still, this didn't stop people from calling her a cross-dressing gay, and "Aunt Janny." "The culture is confused," wrote Mundy, in a meditation on the obsession with Reno's gender and sexuality. "Is she Ma Kettle, hillbilly dominatrix? Is she Nurse Ratched, humorless and repressed? She's hit from all sides. She's a man, she's a woman, she's a man-woman, she's straight, she's gay—but amazingly often, she's depicted sexually."

"Janet Reno's Dance Party" tackled Reno's sex life head-on, but, to Mundy's point, vacillated between portraying her as gay and straight, depending on which seemed funnier. In a January 1997 sketch, Reno pulls Secretary of Health and Human Services Donna Shalala (Kevin Spacey, also in a dress) close to her for a slow dance. Reno is the butch, as it were, to Shalala's "beautiful china doll on a shelf."

By October that same year, the mood has changed and Reno is straight, if sexually repressed, and secretly in love with her boss. She wears a frilly pink robe and slippers while lounging in a girlish bedroom. She cuddles and kisses a large stuffed lion doll, imagining it is the president, and fashions a veil for their nuptials out of her drapes. How did Reno feel about the *SNL* depiction of her, and the criticism of the way she looked? "She was thick-skinned," Gorelick recalled. "But you couldn't ignore it."

At the time, Ferrell said he did Reno a favor by playing her as straight, since she was more often perceived as "almost asexual, in a way." Writing a scene in which Reno longs for a man— rather than women, or nobody—gives her "the benefit of the doubt . . . in that we've chosen to portray her as being repressed and dreaming about—men," he said at the time. Conversely, Ferrell told Mundy that he and his writers would only focus on a male politician's sex life in a sketch if that politician had done something scandalous or wrong, like Bill Clinton. Janet Reno's sex life became a tool for comedy because she didn't seem to have one. A powerful, public woman without a husband, children, or rumored romantic prospects was startling. America didn't understand her, so the culture mocked her. Ferrell said he wouldn't have crafted such a sketch if Reno was a "normal woman." "Madeleine Albright, a short little, quote 'normal' woman . . . I don't know if we . . . It's weird. I hate to break it down into something as simple as the fact that she's tall, but it's almost as simple as that," he explained.

It wasn't that simple. "Will Ferrell playing Janet Reno is incendiary given the challenges for women at that time not coming off as masculine when they were successful," comedy expert Kohen said. "Because that dynamic existed in real life, it certainly complicated matters for women." Ferrell said it best himself: Men on *SNL* are mocked if they do something wrong. For a woman, being tall is enough.

Reno took it like a man—a pretty famous one. "Her model was Lincoln," Gorelick said. "If you looked at the things that were said about him. That he looked like a monkey, for starters. He just let it roll off of him and that's what she did." She wasn't just a good sport about the teasing. She participated in her own skewering when she appeared on *Saturday Night Live* as she departed the attorney general's office in 2001. Reno ran for Florida governor in 2002, but was unsuccessful, and soon after faded from public life. She died the day before the 2016 presidential election from complications related to late-stage Parkinson's disease.

THE AUDITION

In July of 1996, UN ambassador Madeleine Albright visited Prague with Hillary Clinton. Some reports from their trip: The women slip out of their hotel, as if on an illicit errand, and stroll through Old Town Square in smart pantsuits. They pause in Wenceslas Square—the site of protests that ignited the Velvet Revolution and the dissolution of communism in the Czech Republic—and window-shop nearby. The women giggle when the wind blows up their umbrellas. They walk in the rain rather than ride in the car, perhaps so they can be photographed being carefree by foreign press. "They talk about their children, the professors they had at Wellesley College, walking, eating and vitamins," according to one account. The pair nibble dumplings and cabbage and "are clearly having a good time." The travels of a UN ambassador and the First Lady of the United States were couched as a girls' weekend for two vapid coeds. But on this getaway, Clinton had political tricks up her smart pantsuit sleeve. She was on official White House business, vetting Albright for a cabinet position. And while Hillary Clinton was unlikable for trying so hard, Madeleine Albright was considered too hard for failing to be diplomatic.

Certain administration officials thought the trip was "so blatantly political that it is dismissed in just two words: the audition." Albright was auditioning for the post of secretary of state, and Hillary Clinton, who would later hold the post herself, was the casting agent. Clinton was sent to determine whether to anoint Albright as the first woman for the job. Some were unnerved that the First Lady would command such influence, while others felt their jaunt was outright silly. Former secretaries of state like Warren Christopher and Henry Kissinger hadn't needed to audition. They won the role based on diplomatic prowess and political acumen. To frame Albright's bid for the job in theatrical terms, and to describe the trip as a dining and shopping getaway, diminished both the office and the diplomat.

Albright's political career began in the early 1970s when she was forty-five years old, after she had supported her newspaperman husband's career and given birth to three girls. She had earned both a master's and a PhD from Columbia, and taught international affairs at Georgetown, winning Teacher of the Year four times. Albright hobnobbed with Washington elite, often hosting salons at her home. She worked for Democratic senator Edmund Muskie, was recruited to the National Security Council, and advised Michael Dukakis's 1988 presidential campaign.

From the moment she appeared on the national and international stage, she was subject to a chorus of criticism. As with Clinton and Reno, the slights against her had broad reach, from her competence and intellect to her looks and likability. Albright was fifty-five when she earned her first diplomatic post. In January 1993, Bill Clinton named her US ambassador to the United Nations, a position she would hold for three years. It was where she would develop her reputation for being hard and even nasty.

Her detractors were quick to attribute her UN appointment to her connections rather than her credentials. They pointed to her wealth and the access she had to Democratic heavyweights like Geraldine Ferraro and Dukakis. Critics called her "an intellectual lightweight," and argued that she had used her prolific networking skills to wheedle her way into the job—as if bringing powerful people together was not a credential for a job as a diplomat. These so-called extracurricular activities were said to deplete her time at the United Nations, "prompting criticism there that she does not take her job seriously enough," reported the *New York Times*.

KUKO AND COJONES

Albright was the only woman on the fifteen-member Security Council and didn't regularly kowtow to her male colleagues, so she was characterized as abrasive, cold, and out of her depth. She "sneered," "sputtered," and "snarled" to make her point, critics said. Albright was a popular television talking head—especially on CNN, which she jokingly called the sixteenth member of the Security Council—because she spoke in sharp sound bites, often lashing political enemies in ways her male colleagues wouldn't dare. When Albright moved to block UN Secretary-General Boutros Boutros-Ghali from serving a second term in 1996, fellow diplomats found her "too confrontational," and deemed the ouster she led "monstrously handled," despite all the support she received for the ballsy endeavor. Lawrence Eagleburger, the secretary of state under George H. W. Bush, lambasted Albright over the Boutros-Ghali incident in the press. "She is like a bull-dog who gets its teeth into the bone and won't let go," he said, evoking the "bitch" slur without being explicit.

Not all of her critics held their tongues. In March of 1996, Ambassador Albright toured the wrecked Croatian city of Vuko-var. A Serbian mob recognized Albright immediately. The angry horde grew violent, throwing stones and chanting "Kuko! Kuko!" as she walked with her aides through the streets. Albright didn't cower; rather, she demanded that her staff walk through the debris, heads held high. They were American emissaries, after all. By that time, Albright had become a widely identifiable diplomat. The *Washington Post* explained: "There are very few other members of Clinton's Cabinet who would immediately be recognized by a crowd of angry Serbs in an obscure Balkan town and attacked with stones." In addition to her fluency in English, French, Russian, and Czech, Albright knew enough Serbian to know what they were calling her: bitch.

This was essentially the same criticism of her a month prior, in February of 1996, after Cuban military pilots shot down civilian planes leaving Miami. The pilots gloated that they had "taken out the cojones" of their victims. "This is not cojones, this is cowardice," Albright shot back, inveighing against the act by repeating the Spanish word for testicles. But that word, uttered by a woman ambassador, was unacceptable to her cohort at the United Nations, many of whom were "outraged." A former representative from Venezuela accused her of appropriating "a man's word" that he wouldn't say, "even on my farm." The incident "may have been her worst moment at the United Nations," listed among her many "barbed verbal jabs" that won her enemies, according to CNN talking heads. It proved that she could be "too strident," not a coalition builder or a smooth talker who could achieve consensus.

In the press, as in diplomatic circles, Albright was hectored for not being as well liked as some of her peers. While other beloved diplomats had nicknames like "the magician," Albright was called "the queen of mean." Whether it was negotiating with Balkan states or the Middle East, Albright wasn't smooth or decisive. She "nudged" parties for months toward compromise or "lectured" them about history they already knew. She looked askance at partner nations and was called "crisp," "impatient," and "cold." Former president of the Council on Foreign Relations Leslie Gelb called the treatment of Albright what it was: "sexism." "It is hard for a woman, and it's particularly hard to be the first," he said.

Sexism it was, and barely concealed at that. When Albright flew to Moscow to discuss a NATO bomb threat with the Russian foreign minister, he picked a silk flower off of their table and handed it to her. During another diplomatic mission near Paris, foreign colleagues working late at night confused Albright with a cleaning woman and tried to send her away. Albright reportedly hated press coverage of her appearance, and was then called "thin-skinned" and "frustrated" for saying so.

THE GRANDMOTHER

In an echo of the O. J. trial, where Marcia Clark was the only lawyer called by her first name, Albright was routinely called Madeleine by members of the foreign policy community, even those who didn't know her personally. This contrasted with her all-male predecessors, who were called "Secretary" or their surnames. Richard Holbrooke, a fellow diplomat who served as UN ambassador when Albright was secretary of state, repeatedly called her Madeleine throughout a 1999 interview with the *Washington Post*, some believe to undermine her credibility. "Vietnam was known as McNamara's war," wrote the *New Republic*, referencing former secretary of state Robert McNamara. "Kosovo is known as 'Madeleine's.'" (The article was referring to the ethnic cleansing and subsequent war in the Serbian province of Kosovo.) Women had spent decades trying to rise from the secretary title. It was ironic that perhaps the world's most powerful secretary—who actually wanted to be called by that word—was denied it.

"Madeleine's War" was also the headline of the *Time* magazine story about the Kosovo conflict, and whether Albright was equipped to manage the United States' role in it. Her critics were said to "see Madeleine's War as the latest example of an incoherent foreign policy driven by moral impulses and mushy sentiments, one that hectors and scolds other nations to obey our sanctimonious dictates and ineffectively bombs or sanctions them if they don't." This furthered the notion in Washington that Albright was overeager and underqualified, and, in case you forgot, a woman.

Albright used her grandmotherly persona to her advantage in negotiations with tough parties. She made a game of diplomacy by wearing witty pins to difficult meetings. After an Iraqi newspaper called her a serpent for criticizing Saddam Hussein, she donned a snake pin. She sported a bug pin at the Kremlin after a Russian wiretap was uncovered at the United Nations. She also deployed animal brooches to signal how negotiations were progressing—a turtle if things were slow, a bee or crab if they were heated. To confer with the Russian foreign minister on an antiballistics treaty, she had just the thing: an inceptor missile pin. She made headlines for exercising "jewelry" or "brooch" diplomacy, by calling attention to world matters with her impressive pin collection. And the mostly male officers and dignitaries she dealt with appreciated the good humor her pins indicated. They became conversation starters, and world leaders tried to guess what she might wear next, and what it might mean. They also softened her edges.

SPECIAL PLACE IN HELL

At a 2016 presidential campaign event for Hillary Clinton, Albright was flogged again for a one-liner. "There's a special place in hell for women who don't help each other," she said. It's an epigram she has used for the better part of forty years, so famous and devoid of conflict, in fact, that it's printed on a Starbucks coffee cup. But young women called the remark, which was intended to rally their support, "startling and offensive." Albright was swiftly criticized for "rebuking" and "reprimanding and ridiculing" young women.

The media seized on the malodorous scent of an intergenerational catfight, and the ground Albright helped gain for women seemed crushed beneath her ten-second sound bite. The scorching lines Albright had been known for in the 90s, which resulted in her being called harsh and undiplomatic, dogged her once again. But this time, Albright dealt with the flak differently. Rather than steamroll forward, she apologized. "One might assume I know better than to tell a large number of women to go to hell," she wrote in a *New York Times* op-ed titled "My Undiplomatic Moment." It was her version of pastelizing, or handing out cookies. This event, and the public response, indicates that less has changed in the decades since Albright became the first female secretary of state than we'd like to believe.

7

FEMALE ANGER

Female anger in the 90s was demonized and feared, particularly when it was revealed in the public personas and cutting-edge work of artists and entertainers. For musicians like Paula Cole, Lisa "Left Eye" Lopes, and actress Shannen Doherty, their volatility created "bad girl" personas, which colored their reviews, alienated would-be fans, and led to their bitchification. They possessed fiery tempers and misbehaved in public. Society and the reviewers who covered them reacted by shaming their behavior, belittling them, and reducing them to their sexual function.

PRIMAL SCREAM THERAPY

"I didn't write this song for *Dawson's Creek*," says Paula Cole, seated at a piano. Behind her a glass wall reveals a rock-carved inlet of the Atlantic Ocean peppered with sailboats. The audience laughs, knowing they will soon hear the familiar, earnest "do-do-do-do" that punctuated the opening credits of the popular teen drama that premiered in 1998.

Cole wrote "I Don't Want to Wait" for her grandfather, as he was nearing death. She grew up in the small, puritan New England village of Rockport, Massachusetts, where she is performing tonight. She returned to the area a mother, a Grammy winner, and an artist who wrote a song so popular she was able to live off of its royalties for nearly a decade as a single parent.

"I never in a million years expected *Dawson's Creek* to be so popular," she told me. "It really usurped my career. I'm also grateful, because it gave me years with my daughter. I can't speak badly of it when it helped our lives so much." She returned home in 2008 and took an eight-year hiatus from touring. Cole extricated herself from major record labels and now funds her new releases, like the 2017 collection of jazz and folks songs, *Ballads*, through Kickstarter. She recently let her hair return to gray, having grown tired of dyeing it for years.

In the 90s, Cole's rebelliousness fueled her award-winning music and led to her national bitchification. She was once so infamous that Jay Leno created a Paula Cole doll with tufts of underarm hair to mock her. The presence of pit hair defied industry standards for female musicians. "I didn't expect such negativity and frothing hatefulness around such a thing," Cole said. After Cole was called out, she hid her armpit hair under sleeves for a while. Eventually, like Marcia Clark succumbing to pastels, and Hillary Clinton to cookies, she just shaved it. Cole, like Clark and Clinton, wasn't angling to appease critics as much as she wanted to eliminate the distraction so she could get back to work. "I was just sick of it being something that I did pay attention to," Cole says.

Fame and a music career in the 90s were not what Cole had imagined. "I was very unhappy at the height of popularity with *This Fire*," her breakout album, she told me. "I was cut off from my family and I was working all the time." Cole suffered from anxiety and depression, and said the rigors of the star machine and her "quick ascent" left her with PTSD. "I got my ass kicked," she said. Cole's mother feared she was suicidal.

Some of Cole's unhappiness stemmed from the textbook manipulations she was subjected to as a young woman artist. Because she came up during "a newly progressive time" when "women play electric guitars and openly spew their feelings," as *Entertainment Weekly* put it, Cole was dismissed as just another feelings spewer. Coming on the heels of female performers like Alanis Morissette (whom critics called "the princess of post-adolescent feminist angst") and Courtney Love, Cole was mashed together with them, her songs called either "little-girl type" or "riot grrrl variety." Like her contemporaries, Cole's performances, lyrics, and image faced gender-based scrutiny. She was all too often critiqued not as an artist, but as a girl. "Flaunting her bare midriff and pierced nose, she roams the stage wildly, a tough, strident little girl," the *New York Times* observed in 1997.

Cole recorded most of *This Fire* in a day and a half with Jay Bellerose, the drummer she has played with since she was nineteen years old. The album debuted in October 1996 and her hit "Where Have All The Cowboys Gone" reached number eight on the *Billboard* Hot 100. Cole won a Grammy for Best New Artist in 1998, and she is still among only a handful of women whose self-produced albums have been Grammy-nominated.

"Cowboy" satirizes American machismo and conventional domesticity with lyrics like "I will do laundry if you pay all the bills / I will wash the dishes while you go have a beer." But this was the time of political correctness, and the song's irony wasn't initially absorbed. *Spin* called her the "Nancy Reagan of Lilith Fair." The *New York Times* speculated that the lyrics "praise the virtues of docility and domesticity," and could place her "to the right of Tammy Wynette, creator of the memorable retro anthem, 'Stand by Your Man,' as far as sexual enlightenment goes."

Because Cole sang about anger and emotion, she was frequently subject to the "woman scorned" critique. "In the aftermath of Alanis, the airwaves were crawling with troubled ingénues singing tragic ballads about their haunted eyes," wrote *Rolling Stone.*

"Some things come off like, 'This is my journal, there's no self-censorship,'" Rich Silverstein, a musical theater director and Cole fan, told me after the Rockport show. This emotionality draws many fans to Cole, but it made her a critical target. Reviewers didn't seem to know how to discuss a woman artist like Cole without labeling her dramatic, crazy, angry, or all of the above. Her "vocal flutter-kicks and strident self-revelations can be excessive," wrote *Newsweek.*

Reviewers described her performances as primal, self-indulgent, and out of control. She was a "twirling banshee . . . roaming the stage wildly" and found "grand drama in private dilemmas." Like a biblical witch, Cole "unleashed the hellish fury of a woman scorned, bluesily moaned her lust and overindulged in self-administered primal-scream therapy."

Cole's cause wasn't helped by her album's content, or its cover. *This Fire* is, at its core, an album of female rebellion and anger. I loved it growing up because it gave me permission to feel rage that I hadn't known existed in me. It was also sensual and powerful. This was female desire and sexuality expressed and owned by a woman. In the cut "Feelin' Love," for example, Cole sings, "You make me feel like the Amazon running / Between my thighs." The notes she sings sound like orgasm in progress.

The album cover features Cole nude on a swing sailing through blue sky and flames. As a teenager, I was startled to see such a thing. I was used to consuming a steady diet of girls singing about sex in the ways male producers and executives ordained. When Cole did it, it felt different because she was singing for herself and not for men. Shot on a swing set in a Brooklyn backyard, the image on *This Fire*'s cover is about expression, Cole says now. "I may be nude, but it's about freedom and sexiness as reclamation. I'm claiming my body for this purpose," she told me. It's also a taunt at a music industry that quantifies women musicians' value based on the extent to which they're willing to remove their clothes. It mocks the oversexualized album cover, deflating its power. "Here is all of me, my way," it says.

But not all photo shoots would go this way. "I learned to hate photo shoots," Cole told me. "I'd be busy on a touring schedule and I'd come to a shoot and they would show me what I was going to wear, and it would be a little pile of gauze strips," she said. "*Entertainment Weekly* magazine had a little Girl Scout costume. The photographer had this whole plan. And I went along with it and I deeply regretted it."

In a joint magazine interview with other Lilith Fair artists in 1998, Cole recalls wearing a "little pile of rags . . . lying on a table filled with cheese and fruit, like 'You can eat me, too' . . . If you look in my eyes, you can tell I'm not there. The photographer asked me to expose my breast, so I did, and I now want to kill him . . . If you ever, ever want to say no, say no!"

ANGRY WHITE FEMALE

Cole was an "angry woman rocker," and this kind of rebellion was precisely what made the Lilith Fair music tour she performed in so detested. It featured an all-woman lineup to protest how few women acts shared concert billings together, and how radio DJs resisted playing women artists back to back. Lilith Fair was one of the most successful traveling music tours in 1997, grossing $16.5 million. Artists donated a portion of proceeds to women's shelters.

Despite Lilith Fair's visibility and success, it was derided as emotional and unserious. Detractors called Lilith Fair "breast-fest," "girlapalooza," and "the mamas and the mamas." There were plenty of industry folks and concertgoers who expressed excitement about this welcome venue for female artists. But the performance was ultimately immortalized as uncool because it offered "naked good intentions and unbridled sincerity." It was disparaged as "touchy-feely," full of "airy-fairy hoo-ha," and "a slumber party in the woods."

"Call us insensitive, but when we first heard about Lilith Fair we had one reaction: run," jabbed *Newsweek*. "This isn't entertainment—it's therapy." Prescribing therapy for women who expressed emotions was a common gibe hurled by the media.

The "angry woman rocker" label that stuck to Cole, Morissette, and others may have originated with the Riot Grrrl movement. Riot Grrrl began in the early 90s, when women musicians and activists stormed the punk rock stage to espouse feminist values and protest violence against women. Riot Grrrl–affiliated bands—such as Bikini Kill, Bratmobile, and Heavens to Betsy—forcefully reclaimed girlhood and female anger with howling vocals and blatantly political lyrics. The movement was stoked by Bikini Kill front woman Kathleen Hanna's famous demand "Girls to the front" to protect fangirls at their shows from the moshing bodies of men.

Though music was its loudest megaphone, Riot Grrrl was more of a social movement than a musical one. Activists like Hanna wanted less to join a band than to disseminate a message—that girlhood was under attack, and that it needed to be taken back *by* girls. Bikini Kill drummer Tobi Vail put it bluntly: "Not only do we live in a totally fucked-up patriarchal society run by white men who don't represent our interests at all, but we are in a country where those people don't care if we live or die. And that's pretty scary."

For Rebecca Odes, who studied art at Vassar and played bass in the band Love Child, Riot Grrrl was a more accessible form of dissent than sign-waving at protests. "It was basically making art around our anger. It was challenging in a lot of ways popular music definitely wasn't. It was protest music," she told me. The first song she wrote, "Asking for It," dealt with street harassment.

While Riot Grrrl's wellspring was the punk scene of the Pacific Northwest, adherents brought the movement's ideas nearer to the seat of power when they moved to Washington, DC. Riot Grrrls made "fanzines"—handmade publications voicing their ideas that they sold at shows. This was where "The Riot Grrrl Manifesto" was published to outline the movement's grievances and define its goals.

BECAUSE in every form of media I see us/myself slapped, decapitated, laughed at, objectified, raped, trivialized, pushed, ignored, stereotyped, kicked, scorned, molested, silenced, invalidated, knifed, shot, choked and killed.

BECAUSE I am tired of these things happening to me; I'm not a fuck toy. I'm not a punching bag. I'm not a joke.

BECAUSE every time we pick up a pen, or an instrument, or get anything done, we are creating the revolution. We ARE the revolution.

Riot Grrrls attempted to celebrate and normalize female anger. They organized consciousness-raising gatherings across the country. They protested in the name of simpatico causes, like storming Capitol Hill to fight abortion restrictions, and speaking against sexual abuse and eating disorders. They wrote "slut," "rape," and "whore" on their stomachs at performances and protests, and hoisted signs that read "Keep Your Fist Outta My Cunt" and "We Are Not Things." They boycotted mainstream record labels and cultural influences, ate vegetarian diets, and avoided alcohol and drugs.

Sara Marcus, who would later write the Riot Grrrl's history, *Girls to the Front*, was inspired to join the movement by a report about it in *Newsweek*. "It felt like finally jumping into a lake that's exactly the right temperature for your body," she told me. "You would come into a room and there would be an instant sense of recognition. On a cellular level we were made of the same stuff."

Since Riot Grrrl was wrapped in the twin discomforts of female anger and sexuality, the movement walked a line between a curiosity and a threat. Thus, the news media didn't know what to do with it. Its efforts were either misconstrued or outright undermined in press coverage. Its representatives were defused or simply written off. Riot Grrrl was categorized as "feminist fury," "In-Your-Face-Feminism," and "mean, mad and defiantly underground."

A 1993 *Seventeen* magazine article seemed bent on disabling the movement, accusing it of ugliness, misandry, and having a "militant slant." This was supposedly proved by its look. Riot Grrrls "don't shave and deliberately give each other bad haircuts," the report said. Women who didn't conform to mainstream beauty standards weren't real women, as both Paula Cole and Riot Grrrl learned. Unsurprisingly, after the press pigeonholed them as combat-boot-wearing man-haters, angry rape and incest survivors, former sex workers, and caricatures of girlhood, many Riot Grrrls dodged interviews.

The Riot Grrrls' brash optimism and raw passion were also derided as products of youth and inexperience. The movement itself wasn't particularly cohesive. Girls formed bands or wheat-pasted zines around cities from Olympia to Chicago of their own volition, not by checking in with some central brain trust. The movement was leaderless by design—which made for plenty of complications and factions.

What Riot Grrrl did do was to confer on female anger a real place in the popular culture, which then commercialized it. After Riot Grrrl's demise, the very qualities it fostered and the "greater acknowledgement of female anger, messiness, and fierceness" that Marcus says it wrought were celebrated in mainstream female musicians and in consumer culture. "There was no attention to angry women before there was Riot Grrrl," said

Marcus. After Riot Grrrl, "then there was an attention to angry women."

The movement had effectively fizzled out by 1997, as feminists found new affiliations and causes, and associations with Riot Grrrl became more divisive than cohesive. But the ideology and energy reached into the culture in new and expansive ways. So-called angry women rockers flooded the charts and magazine covers, like Fiona Apple and Meredith Brooks, with her anthem "Bitch." Alanis Morissette's *Rolling Stone* cover in 1995 dubbed her "Angry White Female" ahead of the release of *Jagged Little Pill*. Female rage was selling music and magazines, and minting stars.

Marcus says she once might have written off these "degraded copies" of Riot Grrrl's ethos as "bullshit commodified versions," but now she feels differently. "Some women are never going to go to a Riot Grrrl meeting, or have a zine, but are listening to Bikini Kill. Some women are never going to access Bikini Kill, but they're aware of Hole. Some women aren't aware of Hole, but they're listening to Alanis Morissette. Now we're at a third-generation degraded copy, and at the same time, there's an embrace of anger and obscenity that does something for people. Even the Spice Girls are like a fourth-generation degraded copy from Riot Grrrl. So a lot of the effect happened firsthand, secondhand, thirdhand, fourthhand."

Each generation might be more watered down and commodified, but the idea of Girl Power reached more girls all the same. A large part of the Riot Grrrl movement's legacy is its disappearance into pop culture. What was underground in the early 90s surfaced above ground several years later. As Bitch Media founder Andi Zeisler puts it, "Alternative culture in the 90s was co-opted and made into pop culture." Despite the movement's disintegration and lack of recognition, Riot Grrrl certainly deserves credit for introducing Girl Power into the lexicon.

ANGRY BLACK FEMALE

Unless it could be commoditized, like Alanis Morissette on the cover of *Rolling Stone*, public brashness and anger was unacceptable for women in the 90s, mostly because it was feared. Black women's anger was feared even more. When black women expressed their anger in public, they were often subjected to the twin oppressions of racism and sexism. The devaluation of black womanhood that bell hooks writes about in *Ain't I a Woman* in the early 80s was still thriving in the 90s in the potency of the Sapphire and welfare-queen tropes. It's likely why the tough-talking ladies of *Living Single* who challenged men were eventually silenced. Many black artists and feminists in the 90s were similarly disparaged as angry, domineering, and hypersexual. A prime example is one of the most popular musical acts of the 90s, the R&B trio TLC.

Three black women—Tionne "T-Boz" Watkins, Rozonda "Chilli" Thomas, and Lisa "Left Eye" Lopes—broke out in mainstream 90s music by rebuffing sexual victimhood and owning sexual prowess. Their first album, *Ooooooohhh . . . On the TLC Tip*, represented "a major breakthrough in the expression of black female sexuality," wrote critic Nataki Goodall in 1994 in the *Journal of Negro History*. TLC rejected the "loose, immoral and hypersexual" characterizations often foisted on African American women through the Jezebel and Sapphire tropes while demanding sexual agency. Later, they would dismiss loser dudes unworthy of their beds in "No Scrubs," an avowed feminist anthem. TLC's success seemed to prove that black women could both shirk victimhood and own pleasure.

TLC sang about the struggles the black community faced, including gun violence and HIV/AIDS, which were particularly acute in their hometown of Atlanta. They also did for many what school sex education did not—they normalized sex while de-emphasizing its fetishization.

For starters, like some of their female counterparts, they purposely desexualized themselves by wearing baggy hip-hop garb while performing. This choice minimized attention on their sexuality and their looks, and helped them gain legitimacy in the genre. At the same time, their first hit, "Ain't 2 Proud 2 Beg," combined demands for good sex with a rallying cry to use protection. In the video, the group sings, "I like it when you [kiss] / Both sets of lips," while swaddled in multicolored condoms. Sexual pleasure and safety were linked. "Too many kids think condoms are nasty and vulgar, instead of as something that can save your life," Lopes told the *Washington Post.* "Protection is a priority and this demystifies it." Lopes emphasized the message in her public appearances by strapping a sunshine-yellow condom Captain Hook–style over her right eye. It's hard to imagine now, when free condoms are available at schools and businesses, but during the 90s such a display was highly subversive.

"We're consciously working at being role models," Lopes told the *Toronto Star.* "There aren't enough positive ones out there. Not just for young black women, but for all women. Because you guys are still running everything, still making the rules about what a woman can and cannot do and that's not fair."

It turned out that their messages about safety and independence were also commercially viable. TLC's first album went multiplatinum, and by 1992 the band was touring with huge acts like MC Hammer and Boyz II Men. TLC remains by some accounts the bestselling girl group of all time. The only female pop act with higher worldwide sales to date is the Spice Girls.

But increased popularity seemed to muffle TLC's message. And as their clothes grew more fitted and revealing, reviewers accused the group of sexualizing themselves to sell records. Critics called the condoms a trick to hook listeners. TLC was a ruse—preaching safety to sell sex. After they performed on Arsenio Hall's show, the group told the *Los Angeles Times* that other television shows refused to book them unless they "eliminate the condoms and clean up the lyrics."

The 1999 VH1 *Behind the Music* documentary reinforced the idea that TLC was too hot. "Their seductive outfits, sultry dance moves, and sensual lyrics launched these beautiful women to the very top of hip-hop, soul, and pop," explained the voice-over. On tour, men mocked TLC's safe sex message while hitting on them. "They're trying to test us. . . . They think all this TLC stuff is a gimmick," Thomas said in an interview at the time.

Another common critique was that girlish TLC was too young and daffy to be taken seriously. One report called them the "cartoonish-looking trio." *People* magazine described the plots of their songs as girls "standing up to macho men." Since black women's authority and sexuality were at once confounding and scary, TLC's powers needed to be undermined. The group's longtime manager, Bill Diggins, still calls them "the girls," even though they are now in their forties.

EVIL EYE

It was when the girls got angry that the spectacle truly began. In June of 1994, Lisa "Left Eye" Lopes torched the house where she lived with Atlanta Falcons player Andre Rison after a late-night altercation. Rison told *People* magazine that he had slapped Lopes "not to hurt her, but to calm her. Didn't work. We were inside the house now, and I picked her up and slammed her on the bed and sat on her. I still couldn't control her. So I left." Lopes's lawyer later told the magazine, "Lisa is in fear of her life."

Reports claimed that Lopes started the fire because Rison didn't buy her shoes, suggesting that she was both shallow and unhinged. In retaliation, Lopes reportedly filled a bathtub with Rison's sneakers and set them ablaze. The fire spread, destroying the whole house the couple shared.

"Seems that Left-Eye, police say, turned into Evil-Eye," the *St. Louis Post-Dispatch* observed. There was little mention that the home was where Lopes also lived, and that she, too, had lost countless possessions. She was charged with felony arson and released on $75,000 bail.

Lopes struggled with alcohol, and the couple's fights had turned violent before. Less than a year before the fire, Rison—whose nickname was Bad Moon—was arrested in a Kroger grocery store parking lot in an upscale Atlanta neighborhood for beating Lopes and firing a gun. The charges were dropped, so the press simply described their relationship as "combustive," not abusive.

Reports in the story's aftermath rightly implicate Lopes in starting the fire, but the extent to which they cast Rison as the victim, downplaying his part in the violent relationship, feels biased. Rison had admitted in *People* magazine to hitting her the night of the fire, but that was hardly the first time. In the feature titled "In The Heat of the Night: Andre Rison Still Loves Rapper Lisa Lopes Even Though She's Charged with Torching His House," Rison is characterized as a heartsick puppy, while Lopes comes off as "hot tempered" and a vengeful drunk. The lovesick Rison "cried a lot" and "still loves her," and even contemplated suicide—swerving his motorcycle into the median. "Strong child-woman" Lopes, on the other hand, was getting the "crazy bitch" treatment.

That the relationship had been abusive was lost in much of the fire coverage. The reports instead focused on the spectacle— a bathtub of burning shoes, the smashed cars, a house in flames. They suggest that Lopes lit the fire with little provocation. The severity of the abuse is also downplayed. Perhaps it was Lopes's spunky image, her alcoholism, and her pat refusal to be a victim that robbed her of a more sympathetic media narrative. She was sentenced to time in a halfway house, probation, and a $10,000 fine.

At a TLC concert in 2015, I met a former TLC backup dancer who calls himself KitKat (and now sells merch) who adored Lopes and commended her pluck. "She didn't hold her tongue. She spoke her mind," he told me. Others close to her said she fought back when Rison went after her. Maybe there was little discussion about whether Lopes was a victim because she didn't "act like one." She screamed, she drank, she destroyed Rison's property, and she hit back. These things were all true and they no doubt complicated the depictions of her in the press. Lopes was a messy woman who let her anger show, and society didn't know what to do with her. Besides, partner violence wasn't anyone else's business anyway. At the time, it was still thought to be a private matter.

Lopes herself pointed out this challenge. "It's so backward. Andre is a hero, especially in Atlanta's eyes . . . Forget the fact that I got my butt beat," Lopes told *Vibe* magazine in 1994. "But when they saw my face all messed up, they didn't talk about that the way they talked about the house."

Fire jokes abounded. "No doubt Rison is hoping that this will be the last of Lisa's burning ambitions," quipped the *Daily Record*. "TLC burns up the charts and Lisa burns down the house," cracked VH1. Meanwhile, the local papers reminded readers what a good football player Rison was. "There's no denying Rison's brilliance in play," the *Atlanta Journal-Constitution* praised. His boosters vindicated him. "Andre has some hellraiser in him, but he's no hood, either," a friend said. His problems were "usually only harmful to himself," according to his supporters, and his arrest in the parking lot was considered "atypical," since the charges were later dismissed. Rison was softened into a "mama's boy," but Rison's mother did not hold her tongue about Lopes. "She's either going to jail or a mental institution," she said.

After the fire, a *Vibe* magazine reporter spent time with TLC to determine whether or not they were the crazy bitches everyone said they were. "After almost three days of shopping, eating, talking, and chilling with TLC, I fail to see any visible signs of insane bitchiness," she reported. But Lopes's troubles besmirched the whole group, who became "the number one bad girls of pop." They leaned into this caricature to push their sophomore album, *CrazySexyCool*, released five months after the fire. Each adjective in the album's title was meant to describe a woman in the group, and it's no mystery which one was ascribed to Lopes. The album was a runaway success, anointing TLC the first girl group to sell more than ten million albums. *CrazySexyCool* would remain on the *Billboard* 200 for two years.

Many believe Lopes was the creative and artistic fuel of the trio. She is the only group member with writing credits on the debut album. Her raps have a bouncy, youthful quality that distinguishes the group's sound. Theirs is a blend of hip-hop, soul, and R&B that few girl groups had melded before and none to such explosive commercial success. Her spark is what was absent as TLC has reunited to perform again in recent years. While TLC fans supported the comeback, it's lost on no one that the band is eternally incomplete. Lopes was killed in a car accident in Honduras in 2002.

At a 2015 Madison Square Garden reunion show, the audience is mostly composed of women in their thirties and forties, with some millennials sprinkled in. TLC satisfies fans by playing all of their old favorites, but each song sounds exactly like it did then—no creative variations or stylings. The only new song they play is from their 2013 comeback VH1 biopic. They sing it for less than thirty seconds. "It felt like they were put on pause in the late 90s and they just pushed play," says one attendee.

Rather than hire a singer for Lopes's parts, they pipe her prerecorded vocals into their songs. They dedicate "Waterfalls" to her, the song that marketed her struggles and warned about HIV/AIDS. It contains Lopes's signature rap verse about seeing a rainbow while leaving rehab for alcohol abuse. In the carefully constructed narrative of the band relayed to fans now, writing this verse for this song is the highlight of Lopes's life.

As the remaining two performers sing "Waterfalls," the jumbotron plays the music video of the threesome shimmying atop an empty blue ocean for full nostalgic effect. The costly video boasted jaw-dropping special effects at the time, but now it looks dated. T-Boz and Chilli ask the audience to "light it up" for Left Eye, and the dark arena is suddenly dotted with smartphone flashbulbs. "This is nice," says one audience member of the tribute, "but remember when she burned down that guy's house?"

BLUE-CHIP BITCH

Few characters on television were more hated than the blue-chip bitch of *Beverly Hills, 90210*. Loathing Shannen Doherty—who played Brenda Walsh—became an intense, active pastime for many teenage girls in the 90s. By the show's second season, Brenda is no longer its moral compass, protecting friends against date rape and outing shoplifters. Instead, she goes full entitled teen, deploying snideness and tantrums when she doesn't get what she wants. Her purpose on the show shifts to securing access to boys—namely, Dylan. Toward this end, she sasses her parents, throws jealous fits, and wallows in self-pity.

90210 fans blasted hatred at both the actress and her character, who they conflated during the height of the show's popularity. Brenda was "a monster," "a real horror," and "increasingly petulant," but Doherty the actress got it worse. The *Los Angeles Times* wrote that Doherty had a reputation for "snobbery, hostility and general brattiness." Others called her "abrasive," "flinty," "bitchy," and "selfish." Her first notable film role as a mean girl in the Winona Ryder vehicle *Heathers* further solidified her bitch image.

Doherty was the ultimate difficult woman. *Glamour* magazine named her "Prima Donna of the Year" in 1992. A *Sassy* profile called her "Shannen Doherty, pathetic loser," and she was booed at the Billboard Music Awards. She was reportedly financially unstable, bouncing checks, defaulting on credit cards, and stiffing her landlords. Rumor had it that she washed her hair with Evian.

After Doherty was arrested for misdemeanor battery at a nightclub and for DUI on the road, she became a poster girl for female rage. She allegedly brawled with fellow cast members. Their anonymous snipes appeared in print. One said, "She's the same bitch she always was." Another defended the cast but belittled her: "One bad apple shouldn't spoil the whole bunch."

Brenda hate became a cottage industry. Twenty-something Kerin Morataya organized a national cohort to lobby against the actress called the I Hate Brenda Club. Morataya called Doherty "the type of woman that I strive not to be—quick to judge, inconsiderate and very selfish." She threatened that the club would "not stop until Brenda is off the air." Morataya received at least ten calls a day to the "Brenda Snitch Line," a hotline for talking shit. The club produced and sold out of thousands of copies of the *I Hate Brenda Newsletter*, a "wonderfully nasty tattle sheet full of spurious tales and rude doings," praised the *Baltimore Sun*. The club founders created "I Hate Brenda" T-shirts and bumper stickers, a music album called *Hating Brenda*, and a band to perform the songs. The club received gleeful coverage in many newspapers and magazines, and solidified many *90210* superfans' feelings that the show's protagonist was a real-life bitch.

Doherty laughed off the criticism and called charges against her specious. "Do I look like a girl who would bitch slap somebody?" she asked Howard Stern. She joked about the Evian water rumor—"How do you get your hair totally clean that way?"— but nobody laughed. She didn't seem to get the memo about the importance of female-character likability. "I'm not saying I don't have my moments of bitchiness because everybody has them. But it's never for no reason," she explained. *SNL* spoofed how deeply she was hated in the 1993 sketch "Salem Bitch Trial." Doherty plays a woman sentenced to death for practicing "Bitchcraft." "Why is it when a man speaketh his mind, he's admired and made judge? But when a woman displays forthrightness, she's accused of being a bitch?" she asks. "Your words would sway greatly more had they not been delivered in such a bitchy manner!" retorts the governor in charge of her fate. "You shall be burned!"

Doherty's youth, beauty, and refusal to apologize for outbursts fueled the contempt. Progressives didn't like her Republican leanings, which began as a young Southern Baptist growing up in Memphis, Tennessee. Two ex-fiancés were two too many. Reports emphasized what a bad influence she was on fellow cast member Tori Spelling, who upset her famous father by copying Doherty's "hard-partying" ways. Doherty reportedly started a fistfight with fellow cast member Jennie Garth while on set.

NO SAINT

Like Lopes, Doherty was no saint. Both of her ex-fiancés alleged assault—though one, who was the heir to Max Factor cosmetics, admitted in court to slapping her. In a *People* magazine article focused on Doherty's abusive behavior, her father claimed that her fiancé, Dean Jay Factor, was the one abusing her.

"He initiated the charge, but she's the victim," he said. Factor had filed a domestic violence restraining order against Doherty, claiming she threatened to shoot him and he feared for his life. Partner violence perpetrated by a woman was so surprising that the media didn't know what to make of it. For both Doherty and Lopes, their violent behavior caused the press to villainize them while softening the focus on their significant others. Reckless aggression seemed to have graver consequences for the women than it did for the men.

Projecting the image of a train wreck earned Doherty national condemnation, but by scuttling the likable, fuckable, nice-girl identity, Doherty was exhibiting a kind of power. What critics seemed to want most from Doherty was deference. They wanted her to apologize for being a diva, and to mollify fans.

Self-righteous columns advised Doherty on how to extinguish her track record of horrid press. "Shock your public: Say you're sorry!" and "Lighten up!" they counseled. These critiques reeked of a particular brand of fandom that dovetails with narcissism—that an actress owed her followers a certain kind of likability, and that she should care about what they thought of her personal life. When *Playboy* reportedly offered $300,000 for a Doherty pictorial, the prude police called it an "undistinguished honor" she'd share with trash like La Toya Jackson, Mimi Rogers, and Vanna White. When she did it anyway, Howard Stern shamed her on his show. "Baby, don't get crazy," he chided. "There isn't a guy here that doesn't want to bang you. You can't possibly be insecure," he said, as if her self-worth hinged on her fuckability.

Doherty seemed to not care that people called her horrible names and wanted to sabotage her success. Like Lopes, she was accused of unseemly anger and violence. In this way, she occupied male space by goading her enemies to despise her more. The more people hated Doherty, the bitchier she seemed to get. By not capitulating to scolding or judgment and dismissing her enemies' hate, Doherty wielded a unique kind of agency for Hollywood.

Doherty left *90210* in 1994, and seemed to all but disappear until producer Aaron Spelling cast her in his drama about three sexy witch sisters, *Charmed*, which premiered in 1998. Critics cackled at the perfect casting—her bad-girl status had earned her the villainess role. Only playing a witch could bring back Doherty from what seemed like forced retirement. But she left that show, too, after just three seasons, amid another rumored feud with a cast member.

8

MANLY

Since the 90s, we've learned a lot more about anger, specifically how men are rewarded for it while women are penalized. A 2015 study of gender and anger published in the journal *Law and Human Behavior* found that when men and women made the same arguments, men gained influence when angry, while women's influence diminished. Jessica Salerno, a study author, said anger made men seem "credible," while it made women seem "emotional." This was the case for Doherty, Lopes, Cole, and countless other women who attempted to claim power through anger in the 90s. Powerful women were threatening— angry women most of all—and the culture suppressed them through bitchification.

But some women were so angry, subversive, and threatening that they were not so easily subdued by ordinary bitchification. To stop these women, critics in the media and popular culture performed virtual gender-reassignment surgery—crafting male identities for women who undermined cultural mores about femininity and women's place in society.

Roseanne Barr was one such woman. In fact, you could argue that Roseanne Barr was the most hated woman in America in the early 90s. She was loud, crude, brash, and overweight. She was ambitious, and refused to acknowledge her critics, or take orders from network bosses who had green-lit her hit sitcom, *Roseanne*. Her following was built on the fiction that the actress and character were the same person. A former waitress, ex-prostitute, and mom to three kids, Barr had lived the life the show presented—one with a lot of bad road under her.

Roseanne premiered within a month of *Murphy Brown* in 1988. Since there were so few strong women leads with shows built around them, audiences were forced to pick sides—coiffed careerist or pudgy rugrat wrangler? Both were "tough, outspoken women who suffer neither fools nor male machismo with any trace of grace," and "drawn as extremes" since they were so "cutting edge," declared the *Chicago Tribune*.

Murphy was the aggressor in the office; Roseanne was the beast at home. While Murphy appealed to the elite "highly paid professional woman" whom Dan Quayle had chastised, Roseanne Conner was her working-class, overweight, Midwestern cousin who clipped coupons in frumpy sweats. She was a mash-up of spitfire and clown—crafting balloon animals, then gleefully popping them. Murphy worked to keep it together in a man's world, while Roseanne exposed domestic bliss as anything but.

Bergen and Barr were often both nominated for awards in the same categories. *Murphy Brown* ran for ten years; *Roseanne* for nine, ending in 1997. But while Bergen looked and behaved like the goddess movie star that she is, Barr more often looked and acted like a hot mess. This made her efforts on the show more credible, but also etched a bull's-eye on her back.

What irked Roseanne's critics most wasn't her transgressions as a working mom, but as a woman. For one thing, she was overweight, which America detests in women because it undermines traditional beauty standards and projects a lack of ladylike control. Barr stood for "all that we most loathe," according to commentators, with dirty hair, a potty mouth, and a frame that "unapologetically vacillates between being fat and very fat," professed the *New York Times*. She was "Quasimodo on a bad night, a bowling ball in search of an alley," said one critic in the *Chicago Tribune*. The comic, of course, brilliantly played to these barbs. When a *Guardian* reporter dines with her, Roseanne says she's "trying to overfeed." "Takes a lot of discipline to get 2,500 calories in there no matter how painful it is," she tells him.

Barr rejected femininity, and her shtick pummeled its trappings. That she was so harshly rebuked made it clear that critics still held her to a standard that her show mocked. Barr was said to "make many prominent feminists uneasy . . . because she's not terribly, well, feminine," as if heels and pearls were feminist credentials. But her in-your-face attitude and masculine swagger were exactly the opposite of characteristics that 90s women were encouraged by society and industry to embrace. If strident, militant, bitchy Roseanne was a feminist, then I'll join a different club, thank you very much. Throughout the show, her character's husband, played by John Goodman, calls her the floral "Rosie" without irony. Barr's signature joke remains to this day: People say I'm not feminine. Well, suck my dick.

THE MOST HATED WOMAN IN AMERICA

"Why do you think that everyone thinks that you're the good one and I'm the bad one?" Courtney Love asks her husband, Kurt Cobain, in their Los Angeles apartment in the spring of 1992. The footage appears in the documentary *Kurt Cobain: Montage of Heck.* They are wearing bath towels and seem high. Love says that if Cobain cheated on her, she would gain two hundred pounds and become the most hated woman in America. "You're already the most hated woman in America," he tells her. "You and Roseanne Barr are tied."

Cobain was right; the public detested Love. She was loud, brash, and took up space usually reserved for male rock stars. Her power, ambition, and calculations for greatness were too much.

Before punk rock, Love's life was a tortured jumble: an abusive father who reportedly used pit bulls to discipline her, reform school after shoplifting, stripping in exotic locales like Japan and Taiwan, and "gyrating in G-string and pasties at Jumbo's Clown Room, a mini mall strip joint on Hollywood Boulevard." This itinerant and theatrical backdrop inspired Love's sensibility. She was a well-read feminist, apt to reclaim the word "bitch" for herself or quote Simone de Beauvoir.

Her look was vagrant—"a curdled version of the all-American girl." She favored disheveled baby-doll dresses and smudged puppet makeup, intentionally gutting the culturally celebrated little-girl aesthetic with a look that was called kinder-slut or kinderwhore—and helped seed a feminist, progressive bent among one of the biggest male rock bands at the time. Amy

Finnerty, a former MTV programmer credited with discovering Nirvana, told me that Love, Cobain, and his band proactively built a quality and trustworthy group of people to tour with Nirvana. "I remember the jokes backstage like, 'No rapes on this tour,'" says Finnerty. "No one gets raped on a Nirvana tour. Nobody's handing out backstage passes for blow jobs. That doesn't happen in our world. We don't treat women like that. It was such a *thing* that we would talk and laugh about all the time. It's not funny, but we could see the ridiculousness. Those clichés came out of the 80s and the types of guys who would be on crews."

Love, fronting the band Hole, claimed a place for women in punk rock, but many missed the point. "No wonder Courtney Love behaved like an absolute bitch in her baby doll phase, it's the only way to counterbalance all the frills," wrote the *Evening Standard*. She wasn't thin; rather, she had a "bulky" body, like a "young, gothy Bette Midler," *The Observer* explained. Those who believed she wasn't conventionally pretty accused her of "taking up public space reserved for the traditionally attractive woman," according to a feminist journal. That she deviated from traditional beauty standards threatened women who abided by them, and men who determined them.

THE MALE ROCKER ACT

Love's bravado on stage and volatility in the press suggested she was more akin to male rock gods than female songstresses. Love's "unpredictability" and "element of danger" were qualities "we're not used to seeing in a woman. We're used to seeing that from Jim Morrison, or Iggy Pop, or from Johnny Rotten in the early days of the Sex Pistols," said music columnist Lisa

Robinson. Love straddled the stage, roughhoused the mic, and propped her leg up on amplifiers during guitar solos—all male-rocker gestures that she at first mimicked, then revolutionized. Her frequent crowd-diving was another appropriation of the prototypical dude rocker act.

This bold, masculine style also came to define her sexuality and how she used it in her art. Love's stage performances were bawdy and theatrically raunchy, unabashedly drawing from her strip club past. She'd simulate fellatio on a teenager who found his way on stage, strip to skivvies, and hurl herself into a mosh pit, where she often got groped. She aired the details of her sex life—particularly when it involved famous men like Billy Corgan and Ted Nugent. She discussed experimenting sexually with girls, and bathed naked next to a *Vanity Fair* writer.

No archetype is as admonished in our culture as the bad mother. From Love's reported drug abuse while pregnant to the tortured relationship with her daughter that led courts to grant guardianship of Frances Bean to the girl's grandmother and aunt, Love has long embodied this reviled figure. "How many drug-addled rock stars with children have been taken to task for the same behavior?" asked former rock critic Kim France. "Not Keith Richards; not John Lennon; not Aerosmith's Steven Tyler, who didn't even acknowledge his daughter Liv Tyler until she was in her teens. Leaving behind a trail of babies conceived with groupies has long been a key to guy-rock mystique." But for women, like Love, bad parenting was cause for censure.

Another masculine quality Love displayed was what the *New York Post* called her "notoriously vengeful" nature, which she mostly trained on the media and journalists who shaped her public persona. She threatened to kill *Vanity Fair* writer Lynn Hirschberg with Quentin Tarantino's Oscar after the magazine printed that Love used heroin while pregnant. As for the magazine's editor-in-chief, Tina Brown, Love quipped, "I have lurid sexual fantasies about Tina Brown. I'd like to tie her up and run her over with a car."

Love crashed a 1995 MTV interview to talk trash to reporter Tabitha Soren. "She felt like I had some beef against her when I talked about her on MTV News," Soren told me. "I didn't ad-lib my opinion. I read the news. It was a feud she dredged up in her head." Musician and Riot Grrrl Kathleen Hanna said Love punched her in the face backstage at Lollapalooza, for reasons she still doesn't understand. Love later said her gynecologist told her she "had too much testosterone" and prescribed her more estrogen by way of birth control pills. "She could have held back and presented herself with more decorum. But that's not who she is. She's raw," the director Brett Morgen, who made the Nirvana film *Kurt Cobain: Montage of Heck*, told me. "She lets herself go there. And that's rock and roll."

Though Love acted like male rockers, she was treated differently because she was a woman. Her band Hole's song "Asking for It" was inspired by a stage-diving incident during which Love's underwear was torn off and she was assaulted by the rabid crowd. "I can't compare it to rape because it isn't the same. But in a way it was. I was raped by an audience—figuratively, literally, and yet, was I asking for it?" she wondered. This incident earned her little pity. She was often called a "witch," "vampire," or "monster," but never a victim of sexual assault.

Love's original sin may have been marrying and procreating

with rock god Kurt Cobain. In response to an issue of *Sassy* magazine that the couple covered, readers called her a "vile hag" who was "so obnoxious and pushy that she scared Kurt into going out with her." The *Washington Post* said the saddest part of Cobain's tragic tale is that "his talent and his wounded sweetness gets drowned out by the shrill and grandstanding behavior of his monstrous wife."

Love also physically dwarfed Cobain—"three inches taller than he was, and stronger," according to his biographer. Love had long, lean muscles, while Cobain looked frail in his fuzzy cardigans and stringy hair that he washed once a week. Love's personality was also larger than Cobain's. The rumpled-slacker love interest was an icon peddled to women in the 90s—from Ethan Hawke in *Reality Bites* to John Cusack in *Say Anything*. They lacked ambition and were hard to read, but women imagined they contained depths worth probing. They might not have jobs or direction, but they were in touch with emotion and would love you for who you are, or so the script went. Cobain drew many fans, particularly women, with this "new man" shtick. But there wasn't an affable "new woman" model for Love to inhabit. In many ways, their relationship looked like traditional gender role reversal. Those who hated Love attacked her for being "the man," and not a real woman.

After Cobain's suicide, fans accused Love of killing him herself or driving him to it. She still contends with such hatred today—fans blame her for his death in comment threads of nearly every Instagram photo she posts. Below one image of the couple with their infant daughter a commenter wrote, "Fuck you Courtney. Be a real woman and just admit the truth. We miss him and hate you."

RAGE IS HER OZONE

Roseanne Barr was not known for carrying a tune, so it was something of a surprise when she was invited to perform the national anthem at a San Diego Padres baseball game in June of 1990. Wearing a man's shirt and a Cheshire cat grin, she steps to the microphone and asks when she can start.

"Right now?" she asks, seemingly in character as the daffy stentorian Roseanne Conner. She speeds through the open and drops the key a couple of times. She laughs in the middle of a stanza, but when she winds up to "the rocket's red glare," she screeches, and that's when the boos erupt and practically drown out the rest. Her version has since topped lists of all-time worst renditions of the national anthem. (Yes, they make those.) But what incensed fans most was her sign-off. Barr spits on the field and grabs her crotch with exaggerated flair. She could piss on the home front, the economy, even lazy husbands, but America's pastime was sacred. Ticket holders called it a disgrace. Utter contempt followed.

Moral outrage trumped satire here, and the angry mob didn't seem to care that Roseanne Barr, a famous comedian, was joking. Her "denigration of the anthem" was "obscene" and "disgusting." Condemnations included that her rendition sounded "as pleasing to the ear as a fingernail scratching a blackboard," and that it warranted the many boos for the "obnoxious pig" and "ultimate vulgarian." Opera star Robert Merrill, who had spent eighteen years singing the anthem at Yankee Stadium, called Barr's "a national disgrace" and likened it to "burning the flag." A spokesman for veterans claimed Barr had blemished America.

President George H. W. Bush even joined the chorus, regarding the act as "disgraceful." Headlines were irresistibly catty and many used combinations of "fat lady" and "singing" to make their point.

Barr often volleyed between "crazy bitch" and "genius eccentric" in the eyes of the public, as well as her cast and crew. "Not since the '50s has one woman so dominated television," oozed a January 1992 issue of *TV Guide*, featuring Roseanne on the cover dressed as a rogue Lucille Ball with orange hair and a manic expression. She hosted *Saturday Night Live* and mocked her own overexposure. But she was unwilling to cede creative control of her work. Male auteurs who do this are considered brilliant, but micromanagement earned Roseanne a reputation for being impossible to work with. Her sitcom ran through six executive producers during its nine-year tenure. When *Variety* asked one what he would do next, he said he intended to "vacation in the relative peace and quiet of Beirut." A 2016 study showed that women who lead like men are perceived as "bossy, and less effective than male counterparts who behave the exact same way." Bossy Barr became infamous for firing agents, attorneys, assistants, and others who didn't bend to her will.

By Barr's own account, her show's first season was stifled by the network executives and directors who exerted control and stole credit for her work. When the show hit number one in December of 1988, she fired the people who crossed her. Their names appeared on a running hit list she kept on the back of her dressing room door. ABC sent her a chocolate "#1" to celebrate. "Guess they figured that would keep the fat lady happy—or maybe they thought I hadn't heard (along with the *world*) that male stars with No. 1 shows were given Bentleys and Porsches," she later wrote. She asked George Clooney, who appeared in the first season as Roseanne's boss, to hit the chocolate with a baseball bat, and Barr snapped a photo and sent it to the higher-ups. When producers of her cartoon show, *Little Rosey*, demanded she add more male characters, instead of acquiescing she quit after one season. "Rage is Roseanne's ozone. She creates it, she exudes it," *New Yorker* theater critic John Lahr wrote.

Over time, *Roseanne* began to lose viewers. In response, when she tried to evolve the role she'd played for decades—the crude mocker of the domestic goddess and champion of white trash women—detractors dubbed her an inauthentic "prima donna" and fame whore. An acquaintance told *Slate* the show was "the only good idea she's ever had." Detractors claimed that fame had made Roseanne grotesque and unable to connect with her working-class fans. She was also called a fraud for not actually being the working-class woman she played on TV. Barr's qualities that fans had found refreshing—her bawdy, take-no-prisoners humor and lack of self-censorship—were now being criticized.

In 2012, Roseanne made a bid for the Green Party nomination for president of the United States, claiming to speak for "workers, mothers, decent fucking human beings." The 2015 documentary *Roseanne for President!* followed her on the trail. While many thought her campaign was a joke, she was serious. "I have bigger balls than anyone," she says in the film. "I'll say anything to those who need to hear it." Anyone familiar with Roseanne knew this was no lie. "Women still take her more seriously," Eric Weinrib, the film's director, told me. "Men laugh her off."

TOXIC AMBITION

Like Roseanne, Courtney Love was often called trashy, particularly in the face of tragedy. After Kurt Cobain's death, detractors attacked Love for granting MTV News an immediate interview. In it, Kurt Loder prods her, "Did you actually get to say something nice before the end?" She shakes her head no like a scolded child.

Her husband's death also made her ambition attract more bile. "Courtney was interested in the canon and her place in it," Kim France, the former music journalist who covered Love, told me. She didn't pretend to "contain her huge appetites," wrote the *New Yorker*, letting her hard work and desires to achieve show. When she crossed over into Hollywood and wowed audiences with her Golden Globe–nominated performance in *The People vs. Larry Flint*, her ambition was deemed too male, "too mammoth to be confined to a genre," wrote Kylie Murphy in an essay in the journal *Hecate* on Love's relationship to feminism. Her "open ambition is twisted into a pathology that is communicable to everyone with whom she comes into contact," not least of which her husband.

Love excelled at both music and film, boasting "the kind of ambition most people would associate with a male rock star," fellow performer Justine Frischmann told *New York* magazine. Even organized feminism seemed to condemn Love, or at least distance itself from her antics. Love was never one to fall in line with a group or a doctrine. Perhaps as a result, her boundary pushing and rebellions weren't celebrated. "Love's vilification as a bitch is a caution to feminism not to despise blonde ambition,"

Murphy wrote. Love was a "mad, loud, mixed-up, manipulative and unruly woman."

Love's musician ex-boyfriend Rozz Rezabek-Wright called her ambition toxic and even murderous in the 1998 BBC documentary *Kurt & Courtney*. "She thought it was a male-dominated world," he said. "She thought the only way she could achieve stardom was through a man. She had an agenda for me. She wanted to make me into a rock star to the point where I stopped wanting to be a rock star. I wanted to do anything but get away from it. I would've ended up like Kurt. I would've ended up shoving a gun down my throat," he says, blaming Love's drive for Cobain's suicide.

Rezabek-Wright's stinging assessment of Love might have been partially correct. In the 90s, the more public space she seemed to take up, the more hate pelted her. And it was Love's relationship with Cobain that launched her into the national conversation in a new way, garnering the couple the degree of unrelenting press coverage usually reserved for heads of state.

Love was deemed a leech, sucking the star power from Cobain. That she was only famous by association to Cobain was a common slight, but untrue. Cobain biographer Charles R. Cross notes that the two were comparably popular when they met, but that Love was likely ahead in her career. "She knew far more about the music business than he did, and Hole's career was accelerating as quickly as Nirvana's at the time," Cross wrote.

Those who hated her even assumed that *he* was the creative genius behind Hole's breakthrough album, *Live Through This*, which was released the same year he died. "That record was amazing. Everyone tells me Kurt wrote all the songs on it," former MTV News reporter Tabitha Soren told me. "That's

a sexist assumption. I don't want that to be true." As these un-
founded beliefs suggest, Love's fame was considered undeserved,
not only because she hadn't earned it, but also because she had
dared to want it loudly and publicly, like a man.

THE GLOSS

By the late 90s, Love succumbed to the traditional femininity
celebrated in Hollywood. She reportedly lifted her breasts, fixed
her nose, and had liposuction. She turned heads at the Academy Awards in classic red lipstick, a chic blonde bob, and satin
reminiscent of Marilyn Monroe. Love appeared in high fashion
magazines like *Vogue* and *Harper's Bazaar. Spin* interviewed Love
with the cover line "Bitch, Sellout, Murderer," and asked her to
justify modeling for Versace. In one image, she sports men's underwear.

This new Love who had finally capitulated to the mores of
female celebrity was thrown under the bus as promptly as the
old one. Critics instantly longed for the bad girl of yore with
her kinderslut wear. The loss of Love the rebel was mourned. "It
seems a pity that Love has gone the way of the gloss, that she has
tamed the wild child who beat her fists against the straight world
and given us what we surely don't need—another movie star
who's pretty on the outside," wrote journalist Daphne Merkin
in the *New Yorker.* Love had tried to "whitewash her tarnished
image as a trailer-trash junkie," according to the *Washington
Post*, but the nation wasn't buying it. These critiques proved that
no measure of Hollywood styling or rebirth could save Love. She
had sinned too greatly and too publicly—from embodying the
culture's most hated trope, the bad mother, to baldly claiming

the success she wanted and cementing her permanent place as the "black widow" to one of the world's most famous and beloved dead rockers. Love knows this. As she told *Vanity Fair* in 2011: "I thought, like Madonna, I could change my persona with my look. But what I didn't understand . . . is they'll never let you live it down."

Love did the misbehaving, unhinged rebel rock star as well as it's ever been done. "When you say 'rock star,' one of the first things that should come to mind is a picture of Courtney Love wearing a silk slip with one leg up on the amp," said Adam Diehl, an English professor at Augusta University in Georgia, who teaches a course on 90s female rockers. But most people don't think of Love. They think of men—Mick Jagger, Axel Rose, and the like. The same thing could be said of Barr, who defined comedy in the 90s as much as Jerry Seinfeld or Jay Leno. But Barr and Love aren't remembered as artistic geniuses. They're remembered as third-degree bitches.

9

DAMAGED GOODS

Though not new, self-injury gained more attention and close scrutiny in the 90s. Previously, psychologists had thought self-harm was an outcropping of other diseases and disorders, but during the 90s they began to believe that the act itself was worthy of further study. Research on self-harm still doesn't necessarily reflect the full extent of the pathology, since respondents often must self-report, and self-injury behaviors are extremely stigmatized. But adolescents were the population most at risk, and in one 2005 study, 15 percent said they had engaged in self-harm. Boys were more likely to burn or hit themselves; girls tended to cut.

Mary Pipher said she never treated a girl who cut herself for the first decade she practiced psychology, but by the time she wrote *Reviving Ophelia* in 1994, self-mutilation was a common outlet for girls coping with pain. One of her clients, Tammy, covered her bed with newspaper, then cut her breasts with a razor after her boyfriend assaulted her. She began to do it whenever the pair fought. "Self-mutilation may well be a reaction to the stresses of the 1990s," Pipher wrote. If eating disorders spread out of cultural pressures to be thin, then cutting represented women's need to "carve themselves into culturally acceptable pieces." Piercing and tattooing culture were also manifestations of this desire to mutilate the body, she said.

The depression driving these acts was being addressed in private offices like Pipher's, but less so by society at large. As girls' depression climbed, so did their suicide attempts. A 2001 report, *Risky Behavior among Youths*, dedicated an entire chapter to the rise in suicide among fifteen- to twenty-four-year-olds. Using data from 1995, the authors showed that nearly 17 percent of girls said they had considered suicide, while nearly 6 percent had attempted it. More than twice the number of girls attempted suicide than boys, though boys succeeded more frequently. Girls were more likely to make a suicide attempt if they knew someone who had previously done so. Psychologists characterized self-harm as a coping mechanism for handling difficult emotions and a potential marker of low self-esteem and perfectionism that plagued girls in the 1990s. That girls' anger and emotions were not validated, and were often outright mocked, didn't help matters.

While the 1987 book *Bodies under Siege* calls itself the first to explain self-mutilation, it took nearly a decade for the act to lodge in cultural dramatizations. By the late 90s, cutting, as it was colloquially called, was represented and romanticized in pop culture. A 1996 film, *Female Perversions*, dramatized self-harm, as did *Beverly Hills, 90210*. Psychotherapist Steven Levenkron's bestselling novel, *The Luckiest Girl in the World*, centered on a female self-mutilator. *CosmoGirl!* magazine founding editor

Atoosa Rubenstein wrote about cutting herself while she was in college. Cutting was called "the new bulimia."

Perhaps the watershed moment for self-harm was a 1995 interview with Diana, Princess of Wales, in which she admitted to cutting her wrists with a razor and her chest and thighs with a penknife during her unhappy marriage. She also discussed suffering from bulimia. These revelations from royalty shocked the millions who had watched her storybook wedding. In the US, the *New York Times* called cutting "a national obsession." One author who spoke to dozens of cutters for a book called *A Bright Red Scream* referred to self-harm as "the addiction of the 90s."

Breathless headlines and popular culture representations of girls' drug use, eating disorders, sexual promiscuity, and self-injury conspired to create the "damaged girl" or "damaged goods" stereotype. The 90s damaged girl was often young and rail-thin—from drug abuse or disordered eating, or both. Usually, she was introduced to the public as an innocent girl-next-door who lost her way with sex, drugs, or psychological issues. She was thought to be immature—hence the tag "girl"—and overly emotional, and she lost her virginity or was sexualized before her time. Media narratives referred to her as dirty, broken, or in need of saving. Her fallen-ness was an excuse to savage her. The trope was easy to apply to any girl who didn't seem sufficiently obsessed with perfection. The "damaged girl" or "damaged goods" was a motif related to the inculpated victim. She was at once ridiculed for her suffering and fetishized in countless films and television shows watched by 90s girls.

Notably, the damaged-girl trope mostly applied to straight white women. Their struggles with self-injury were studied and documented more than those of women of color and LGBTQ women. Thus, the damaged-goods stereotype in the media and culture came to reflect a limited view of the epidemic. A 2004 study of LGBTQ women who self-injured said they did so to cope, but that their sexuality and identity played a role. "Self-injury can be understood as a coping response that arises within a social context," the journal authors wrote. Gabriela Sandoval writes about women of color cutting themselves at the Ivy League residence hall, where she served as director, in the essay "Cutting through Race and Class: Women of Color and Self-Injury." She explains that Latina students from immigrant families cut themselves because of pressures, but also "as an expression of the disparity between their own and their parents' experiences." Research and media ignored this broader context when they prioritized straight white women who cut—yet another example of how white women were bitchified differently from women of color in the 90s.

A BAD, BAD GIRL

Pop star Fiona Apple released her first album, *Tidal*, in the summer of 1996, when she was nineteen years old. It was a truth bomb with unsettling chord progressions and attention-grabbing lyrics. What sparked controversy around the ingenue was the music video for her hit "Criminal." The disheveled artist writhes on a carpet amid the remnants of a party. She sings, "I've been a bad, bad girl," while sitting in a hot tub, men's legs straddling her birdlike neck. The suggestive video enticed grade school boys to hunt for it on MTV and scandalized the music press.

Critics said Apple epitomized "heroin-chic" and looked "like an underfed Calvin Klein model," like "Kate Moss with songs." The *New Yorker* targeted her "teenager's sense of drama." "Apple has often seemed part banshee and part waif, an unstable cocktail of ferocity and fragility teetering on the edge of detonation," the *Milwaukee Journal Sentinel* judged. Because she was wearing underwear in the video, many assumed she was prowling for sex—"slinking," "flouncing," and "writhing," according to reviews. *Time* named the video among the top ten most controversial videos of all time fifteen years after its release. But the reason that the video is controversial is lost on most. Apple isn't baiting men with sex or addled by drugs. In fact, she isn't playing for the camera at all—she is telling men to fuck off.

Kristin Lieb, an associate professor of marketing at Emerson College and author of the book *Gender, Branding and the Modern Music Industry*, calls the video defiant. "It's very different than the videos you see—I can think of a million—like Katy Perry's 'I Kissed a Girl,' where it's all a performance for the camera, for the male gaze. The way Apple looks at the camera, she is defying, even shaming its gaze."

"Criminal" seemed to infantilize and sexualize Apple at once with its "overtones of child porn." But Apple was not a child when she made this video; she was a consenting adult. Still, she was shamed and called a tormented little girl. "Pop's newest star is barely twenty, still lives with her parents and doesn't have a driver's license. So what is Fiona Apple doing half-naked all over MTV?" *Spin* asked. The article went on to describe her behavior as "generally acting like a sexy temperamental teenager, the kind of arty ravished girl you knew in junior high who wrote poems in all lowercase letters." She was shot for the story by fashion photographer Terry Richardson, who has since been accused of sexual harassment and assault.

Thanks to her youth and emoting, Apple's critics also labeled her a narcissist—"a self-obsessed drama queen exploiting her psychic wounds." The *Washington Post* called *Tidal* "a tsunami of adolescent feelings in which Apple revealed far too much of herself" (ironic, considering popular music industry standards require female artists to bare as much skin as possible). Apple was no struggling, troubled youth, her critics charged. Instead, she was a spoiled brat. *Rolling Stone* wondered if press coverage of her "might lead one to believe that Fiona Apple is either a precocious, calculating prodigy or an unbalanced, ungrateful freak."

Critics pathologized her for not smiling, and for looking "emaciated" and "too thin." Ever candid in her interviews, Apple confirmed to *Rolling Stone* that she did indeed have an eating disorder resulting from the trauma of being raped at a young age. "For me, it wasn't about getting thin, it was about getting rid of the bait that was attached to my body. A lot of it came from the self-loathing that came from being raped at the point of developing my voluptuousness. I just thought that if you had a body and if you had anything on you that could be grabbed, it would be grabbed. So I did purposely get rid of it," she told the music magazine in 1998.

Unfortunately, Apple's disclosure that she was a rape survivor didn't stop the press and fans from treating her as a piece of meat. "Imagine if you do have rape or sexual assault in your background and now everybody is focusing on your body and whether you're sexy enough, it has to be a nightmare—and she's one of the only ones who really takes that head-on," Lieb said. Instead of commending Apple for shaming men who leered at her, critics accused her of using her youth and her sex for fame. And perhaps she was. But to say her artistic efforts were largely misunderstood is a gross understatement. Even twenty years later, "Criminal" is cheered as controversial, not for this, but because Apple was an artist who looked like a damaged girl—a child singing in her underwear.

Today, fans hunger for artists to call bullshit on the entertainment industry. To post makeup-free selfies, to shirk Photoshop in modeling contracts, to be a few pounds heavier than perfect—all these acts are applauded like a triple axel. Years later, *Rolling Stone* would call *Tidal*—which sold three million copies upon its release—one of the best albums of the decade. But at the time, it sounded like nothing else on the radio, which was both intriguing and confusing. Apple was so ahead of her time, in so many ways, that not only was she smeared for not falling in line, but her protest went practically unnoticed. "It would probably make her brand Britney Spears big now, but at the time she was called crazy," said Lieb.

The damaged-girl critique of Apple culminated when she received a moonman for Best New Artist at the 1997 MTV Video Music Awards. I watched in real time with tremendous discomfort. Apple used her speech to censure the music industry—the very folks who put the astronaut in her hand. "This world is bullshit," Apple said. "You shouldn't model your life about what you think that we think is cool and what we're wearing and what we're saying and everything. Go with yourself. Go with yourself . . . It's just stupid that I'm in this world but you're all very cool to me."

I was Apple's target demographic, yet I absorbed the critiques of her and parroted them myself. Why didn't she say *thank you* and smile sweetly? She *did* look like the cocaine-snorting starved model who so frequently appeared in my *YM* and *Teen* magazines, selling an androgynous fragrance or diet pills. I couldn't stand that her look was popular, mostly because my shape was the exact opposite—breasts, flesh, and curvy hips that my chemistry teacher once called "childbearing" in front of my class. I was never an Apple fan because I was scared of her and threatened by her intensity. What she was trying to do was completely lost on me. As a teenager, I was sold and bought into what Apple was rejecting—beauty, sexuality, and even personality shaped and policed by men. Apple's attempt to undermine the "perfect girl" aesthetic—shaming the gaze, spilling her guts, and starving the flesh from her frame—threatened the affable, obedient, perfect-girl archetype that my peers and I were trying so hard to mirror. I didn't buy her record. I skipped right past it in my friends' CD changers. When her video came on, I changed the channel. I couldn't look at her body writhing on the carpet. All it took was a few notes of the moaning piano intro to "Criminal" or a glance at the video's sepia tones, and I felt sick. Apple was "damaged goods"—something I longed not to be. She upset me. Now I realize that was exactly the point.

THE FALLEN PRINCESS

ABC *World News Tonight* executive producer Kathy O'Hearn learned from a colleague that Princess Diana had been in a car crash. "They kept saying she hurt her thigh. Right away, it smelled much worse," O'Hearn told me. She sped to work, where she found her colleagues—"all dudes with crossed arms standing around"—trying to decide how to cover the story. Should they cut into regular programming? Wait to make a special report? "I said, 'This is fucking huge. This is a princess and an accident. You don't know how people are attached to her.'" Finally, the staff decided Peter Jennings, the face of the network, should report the story. But he was vacationing in the Hamptons.

"I had to say, 'Peter, I want you to come in. The princess was in a car crash,'" O'Hearn recalled. "He said, 'I don't understand this. Why are we making such a big deal of this? She was not a head of state. She did not command an army. She was not the head of a corporation.'" O'Hearn made her pitch, that Diana was a princess at once relatable and real, and that her struggle mattered to people. She stirred emotions—particularly in women—with her stoicism in the face of disease, humiliation, and rejection. Jennings did the story, of course, but "he was just livid that he had to come in that night," she said.

Obviously, Jennings's instinct was wrong. Diana's death was one of the biggest news stories of the decade, inspiring weeks of wall-to-wall coverage that included countless hours of retrospectives on her life, live coverage of the memorial of littered flowers at Buckingham Palace, and her funeral itself. The fixation on her death was a throwback to the event that launched the global obsession with her royal life: seven hundred fifty million people watched Diana Spencer's wedding to Charles, Prince of Wales, in 1981.

Early stereotypes ascribed to Diana included that she was unsophisticated, girlish, and stupid. Her education halted when she married at twenty years old. Her lack of refinement was proved by her look—she dressed in "frilly modest-maiden dresses" and wore an unflattering "modified bowl haircut." Diana recalled being "always pitched out front" to appease press when she traveled as an ambassador with her husband, which led to intense scrutiny on "my clothes, what I said, what my hair was doing, which is a pretty dull subject," she said at the time. Needless to say, her husband's appearance wasn't given the same treatment.

While the prince's interests were serious, hers were vapid. Charles relished discussing philosophy, and doted on horses and his garden. Diana "adored fancy clothes, listening to pop music on her Walkman and telephone gossip," said one report. When she told a sick child at a London hospital that she was "thick as a plank" to ease his nervousness around her, it became a global story.

But over time, the icon and vacuous girl-next-door became the troubled woman. Reports of Diana's mental illness, bulimia, and suicide attempts emerged around 1992, when journalist Andrew Morton published the biography *Diana*. (Morton covered the princess before Lewinsky.) He charged that the royal family deprived Diana of resources and support to cope with her sickness, constant media scrutiny, and invasions of privacy, while mocking her suffering. Charles once pointed to her plate of food and asked, "Is that going to reappear later?"

In her 1995 soul-baring sit-down interview with Martin Bashir on the British news show *Panorama*, Diana revealed her marital struggles and battles with postpartum depression, bulimia, and suicide in rare detail. These subjects were incredibly taboo at the time and not often discussed in public, let alone by a royal. Many called her a master manipulator of the media, and saw the interview as a ploy for public sympathy. But as the princess explained to Bashir, her honesty sprang from desperation, and because she was "fed up" with being seen as a "basket case." She told Bashir that her suffering was ammunition for the royal family. It "gave everybody a wonderful new label. Diana's unstable. Diana's mentally unbalanced."

Her family and the press seemed to believe that she could control her binging and purging, and that her five alleged suicide attempts were not the product of a disease, but a mind-set she could change. She told Bashir that her family called her an "embarrassment" who was "sick and should be put in a home." Her suicide attempts were described as "half-hearted" or "widely seen as cries for help." The queen believed her bulimia to be "the cause of the marriage problems and not a symptom," and later the reason for the royal couple's split, Diana said. In fact, the princess was part of an epidemic. From 1988 to 1993 bulimia rates tripled in the UK, and kept climbing until 2000.

Meanwhile, perhaps because she intimately understood what it felt like to be stigmatized, Diana was an early preeminent face of the AIDS crisis. She was photographed shaking hands with those diagnosed with HIV in the 80s, when it was still widely believed that any contact could transmit the disease. Back then, politicians in America and England were distancing themselves from HIV/AIDS, if not outright ignoring the epidemic and those suffering. But Diana took her young sons to homeless shelters to meet those dying of AIDS, exhibiting a radical empathy.

Diana was a great force in destigmatizing the disease, but the attention this effort won her displeased the palace. Her work to connect with regular people diverged from previous monarchs' behavior, and the press accused her of using her advocacy work to steal attention from her husband. Charles and Diana were immersed in a "rivalry for public attention and approval" that "continued until her death," reported her obituary in the *New York Times.* She took "highly publicized trips to former war zones like Angola to conduct her high-profile campaign against landmines," as if she were gunning for an Oscar instead of striving to prevent the loss of life and limbs. Diana "posed knowingly on Mediterranean holidays" with her friend Dodi Fayed, who died in the same car wreck that took her own life, in an "apparent effort to show the world that the once-troubled young woman had found personal happiness."

From her engagement at age nineteen to her death at thirty-six, cameras followed Princess Diana wherever she went. "I never know where a lens is going to be," she said. Bashir presses this point in their interview. "Some people would say that in the early years of your marriage you were partly responsible for encouraging the press interest . . . You seemed to enjoy it . . . Do you feel any responsibility for the way the press have behaved toward you?" "I've never encouraged the media," she replies. "Now I can't tolerate it because it's become abusive and it's harassment." The question became more poignant after the wreck in which she died, fleeing tabloid photographers on Paris streets. Though her driver was drunk, the tragedy was symbolic. Even her *New York Times* obituary hinted she might be to blame for the press attention and accused her of manipulating the media.

There was no precedent for the future king of England divorcing the future queen, but when Charles wanted out of his marriage to Diana, he and the royal family used his wife's mental condition as an excuse. Charles's extramarital affair with his now wife Camilla Parker Bowles was an open secret. At first, royal family members around Diana called her anxious and "paranoid and foolish" for believing it. They tried to convince her that she had imagined the affair. Later, the monarchy suggested that Diana's madness drove her husband to cheat on her, and used it as an excuse to expunge her from the royal family. "Do you think that because of the way you behave, that's precluded you effectively from becoming queen?" Bashir asked her in their interview.

While both Charles and Diana admitted to extramarital affairs, hers was seen as more shocking and blemishing to her character. She was amoral, "frail," and "weak spirited" for starting an affair with her horse-riding instructor, James Hewitt. The palace blamed the princess for smearing the family with her affairs—including, scandalously, with powerful Muslim men—and blabbing to the press, even though Prince Charles had also been unfaithful for years and had spoken openly to a biographer about his personal life.

Even years after her death, after countless revelations about her struggles, Diana still can't shake the damaged-goods label. The 2013 biopic *Diana* portrayed the princess as vindictive, wanting to settle the score with the royal family she believed duped her into a loveless marriage at a tender age. The film was roundly panned by Diana's admirers because, according to a review by Nicholas Wapshott in *Newsweek*, it "got too close to the truth." "Vengeance is a dish best served cold," wrote Wapshott, "and Diana delivered it straight from the freezer."

Diana, long since buried, remains a damaged girl. Her narrative—about a fairy tale exposed as a sham, romantic and familial rejection, mental illness, and the loss of and quest for love—was one that cut to the bone. But men, who comprised more than half of journalists in 1998, didn't see it that way. Like

Jennings, some rolled their eyes at Diana coverage over the years because her story seemed unserious. This sentiment was amplified the more "damaged" she became.

The damaged-goods trope—and its components of self-harm, sexualization, and objectification—was destructive to girls in the 90s. Sarah Naomi Shaw, a researcher at Harvard, argued in a 2002 article in the journal *Feminism & Psychology* that the way girls' self-injury was studied and treated in the culture "mimics women's experiences of objectification." Objectification theory posits that girls internalize and try to measure up to an outsider's perspective of themselves. Thus, self-harm doubly objectified adolescent girls in the 90s. Girls cut their flesh after being objectified by patriarchal culture; then patriarchal culture objectified (mostly straight and white) girls for cutting themselves by creating the construct of the damaged girl. It was extremely meta. But one thing the damaged-goods stereotype seemed to ensure was that girls would become troubled women.

10

VICTIMS AND VIOLENCE

One of the most popular, sought-after women on television in the 90s was a new mainstream television character: the Woman in Jeopardy. The "Woman in Jep"—or just "Jep" to her network honcho friends—was victimized or put into life-threatening situations in countless made-for-TV movies. She was a perennial hit and commanded huge ratings. Often, Women in Jep were versions of the "damaged girl," which made it acceptable, somehow, for them to be savaged by their abuser or torturer.

Network executives told the press that Jep films empowered women characters, enabling them to assume qualities they weren't normally given by television writers and producers. Women in Jep could "cajole, demand, infiltrate, investigate and settle scores," all in the name of offing a threatening man-monster, said one magazine. These women didn't need a hero to free them; they could save themselves, and did so in an arsenal of skimpy outfits. Of course, in reality, Jep was a gimmick to sate audiences who wanted to see women both suffer and dole out abuse on TV.

Oftentimes, a Woman in Jep needed to overcome a bum rap, like causing her best friend's demise in *The Woman Who Sinned* on ABC, or sexually abusing her own baby on CBS's *In a Child's Name*. A female pediatrician "faces the loss of her practice, her family and her freedom" when she's accused of murdering an infant patient in *Deadly Medicine* on NBC. The wrongly accused baby abuser on CBS, a professional dental hygienist, is beaten to death by her dentist husband. After a while, Jep films began to resemble a choose your own adventure: Fallen mother/angry babysitter/delicate sorority sister encounters homicidal husband/psychopath brother/sadistic lawn-care professional wielding knife/drill/scary whisper voice.

Networks aired some 250 made-for-TV movies in the 1992 season. A woman was physically or psychologically abused in at least half of them, according to a *Newsweek* report about the rise of Jep, cheekily titled "Whip Me, Beat Me . . . And Give Me Great Ratings." There were murderously jealous ex-husbands, alcoholic fathers, and fiendish sons to fend off. These Sunday night films were some of the networks' best-performing content of the decade. Industry insiders called them a remedy, plugged in when programmers failed to create hit series. And many were supposedly based on real-life scenarios and true crimes. Like the stories of Nicole Brown Simpson, Princess Diana, or Anna Nicole Smith, fictional Women in Jep became scintillating content, the kind 90s audiences hungered for. *Entertainment Weekly* summed up the fervor this way: "Somewhere in America a wanton act of criminal violence is being committed. And somewhere in Hollywood, an agent is trying to lock up the movie rights, a producer is pitching the story to a network, and an actress is praying that the tale involves a lethal weapon, a desperate woman, an Emmy-nominatable breakdown on the witness stand."

Hollywood clamored for Women in Jep scripts because they "played well in the flyover," meaning the states between the East and the West Coasts. "It's a case of serial sexual harassment: a dozen TV movies about women in jeopardy in the month of November alone," explained *Newsweek*. When executives heard pitches, "the first thing many say is 'Where's the jep?' (The second is 'More jep!')," the article continued. If a script was boring, executives might demand that women be chased by a knife-wielding hospital orderly or demonic construction worker to liven things up.

The rise of this genre solidified the idea that women enjoyed watching other women suffer, since they were the films' most frequent consumers. Jep became so prevalent that critics questioned and even mocked the trend. "Women are being beaten, terrorized, abducted and killed at an alarming rate," wrote the *Detroit Free Press*. "It does seem like there's a shocking amount of this," Robert Thompson, a television and popular culture professor at Syracuse University, told the *Columbus Dispatch*. "The old traditions have kicked in again where, in fact, often women are the victim."

Not all women wanted Jep. Actress and Oscar nominee Diane Ladd, who appeared in many TV movies, bemoaned the crappy scripts she was getting. "When I have these meetings with network executives, and most of them are men, it's all women in jeopardy movies . . . it's women being abused by Coke bottles," she said in a television appearance at the time. Ladd speculated that women watched the films less because they enjoyed them than because they *feared* such horrors. *Newsweek* noted how some 95 percent of television movie writers at the time were male.

Women were rarely both violent and empowered on-screen. The 1991 film *Thelma & Louise* was a notable exception and a huge hit. The difference is that it was made by and for women. It featured two strong female leads (Geena Davis and Susan Sarandon) and a female writer (a former music video production assistant named Callie Khouri) who won an Academy Award for Best Original Screenplay. Unsurprisingly, many studios passed on the film. One male executive's response to the script was, "I don't get it. It's two bitches in a car."

Another effect of Jep flicks was that they normalized crimes against women. A panel of psychologists who found that television egregiously misrepresented women also discovered that sexual violence on television, like in Jep, "leads to increased acceptance of rape" and "can instigate antisocial values and behavior," according to a 1992 study.

Meanwhile, television executives were quick to blame audiences for the onset of Jep. "We look to the audience to tell us when they've had enough," the CBS entertainment president said at the time.

THE LONG ISLAND LOLITA

Audiences who devoured Women in Jeopardy fictions were primed to consume such stories when they occurred in real life. Amy Fisher was sixteen years old when she became romantically involved with a married auto mechanic, Joey Buttafuoco. In May 1992, Fisher rang his doorbell, pulled out a .25 caliber Titan semiautomatic pistol, and shot his wife in the head. The bullet nearly killed Mary Jo Buttafuoco, shattering the base of her skull, nicking an artery in her neck, and tearing her eardrum. Fisher was described as a "teen girl psychopath." The incident was "Fatal Attraction . . . teenage style."

Fisher's crime became a national headline with the discovery of a sex tape. Buttafuoco had persuaded Fisher to work for an escort service for extra cash, and there was video proof. Eight days after Fisher was arrested, twenty-eight-year-old Peter De Rosa sold the tape to the tabloid television program *A Current Affair* for $8,000. This launched Fisher to national prominence and tabloid infamy. The fourteen-minute video reportedly showed Fisher servicing her client. That she was underage didn't deter broadcasters. The press simply condemned her for sluttishness. "Anything," she tells her john. "I'm wild. I don't care. I like sex."

The story had the perfect ingredients for a tabloid feast in any decade: an underage girl, sex, prostitution, women competing for a man, and attempted murder. It wasn't so much the violent shooting that propelled media obsession, though they did dwell on the catfight; it was Fisher's filmed prostitution. The media cast her as a selfish, unrepentant tramp. Journalists called her a "pubescent prostitute" and "lingerie temptress" who "wore cutoff jeans that fit her like white on rice."

Newsday took on the persona of a sex partner with the headline "Oh Amy, Oh Amy, Oh Amy." When *New York Post* writer Steve Dunleavy interviewed Fisher's father, he admitted that Fisher had been "vilified in print as a venal, spoiled little bitch" by many journalists, including himself. The headline of the piece was "Lolita's Dad Begs Forgiveness for His Girl." The assistant district attorney prosecuting Fisher called her "shrewd, manipulative, and brazen" because she asked clients to contact her directly rather than go through the escort service she worked for. Her sexuality was Machiavellian.

New York tabloids nicknamed Fisher "the Long Island Lolita," alluding to the pedophile victim in the novel by Vladimir Nabokov. What was rarely mentioned was that her relationship with Buttafuoco was, by definition, statutory rape. Instead, the press romanticized their connection and gave it a poetic cast—an "Oedipal tinge," since Fisher was just a girl, and Buttafuoco was "old enough to be her father," *Time* magazine wrote. The literary quality of the encounter was emphasized, not the illegality and immorality of it, nor that Fisher was a victim of sexual abuse at the hands of a much older man.

Buttafuoco was sentenced to a mere six months of jail time. Fisher bore the blame instead, with the popular press depicting her as a villainess and erotomaniac—the wanton and wild slut. "To call her a seventeen-year-old girl who's living at home is as accurate as calling John Gotti a businessman who lives in New York," the prosecutor said. Fisher pleaded to reduced charges that would put her in prison for nearly seven years. She was diagnosed with severe depression and attempted suicide twice. In her memoir *Amy Fisher: My Story*, Fisher writes that she was raped at thirteen by a worker in her house, and later had an abortion. "So here I was, on the brink of sweet sixteen: a class-cutting, report-card forging, ashamed secret rape victim who'd been sexually abused as a little kid," she wrote.

Fisher had few defenders. One columnist at the *New York Post*, Amy Pagnozzi, took up for Fisher and criticized the media for focusing more on her sexuality than her violent crime. "Amy Fisher will be made to pay as much for her sexuality as for any crime she may have committed, while the men will get off like they always do," she wrote. Fisher was clearly the victim, according to Lesléa Newman, author of the 2009 novel *Jailbait*, which was inspired by the Fisher story. "He's going to say whatever he needs to to get into her pants. She has no self-esteem. So he's going to say, 'I love you. I'll leave my wife for you.' Obviously, she shouldn't have shot Mary Jo. But she's abused and abused and then she was desperate to hold on to what she was promised," Newman told me. The bulk of the media coverage, however, portrayed the teenage victim as a calculating villainess.

Fisher's story was a natural fit for Jep, and she sold it to a production company for $80,000. The Big Three television networks ran made-for-TV movies, which the *New York Times* called the "Amy Fisher Film Festival." The press coverage turned Fisher into a *Melrose Place*–worthy villainess who used sex for power, and now Hollywood was glamorizing her for it. The films were among the most-watched TV movies of the year. Since Fisher was portrayed by attractive young actresses like Drew Barrymore, the message was, "If you become a teenage prostitute and go out and shoot somebody, maybe you too can become a media celebrity," according to the *New York Times*.

Joey Buttafuoco's lawyer told reporters that Fisher could go from being "a $180-a-night prostitute to a $2 million-a-night prostitute" because of her fame. The sketch comedy *In Living Color* spoofed Fisher with a skit in which she offered a seminar called Amy Fisher Bang for Your Bucks. In it, Fisher is too dumb to count the number of reasons to sign up for her own workshop, let alone how much cash she earned. When the Buttafuocos sold their story, they reportedly made roughly triple the amount Fisher did.

Fisher was freed from jail in 1999, but her sexuality was so threatening, all those years later, that it still warranted jokes to defuse it. The *New York Times* called her a "chastened hussy," still emphasizing sex. *Saturday Night Live* imagined that the most satisfying part of Fisher busting out of jail would be to reunite with her cutoffs and see-through halter top. *SNL's* version of Fisher, played by Cheri Oteri, was poised to open up a shop selling "classy tear-away underwear" upon her prison release.

GIRL ON GIRL

Curiously, Fisher's crime occurred in the midst of a disturbing trend—a rise in the female share of juvenile arrests in the 90s. A 2008 report compiled by the Department of Justice found that between 1991 and 2000, arrests of girls increased more than arrests of boys for a variety of offenses. By 2004, girls comprised 30 percent of all juvenile arrests. All told, the female arrest rate had nearly doubled since 1980. The report, *Violence by Teenage Girls: Trends and Context*, analyzed crime reports, longitudinal student surveys, and questionnaires filled out by crime victims to understand why female delinquency changed so markedly in the 90s, while male delinquency did not. The authors cited additional research that found girls were more likely to perpetrate domestic violence than boys and were three times as likely to assault family members.

Why the uptick in girls' violence? The report pointed to broad misconceptions about girls' sexuality—specifically, that the culture's appropriation of girls' sexuality did not track with the reality of girls' sexual lives. Girls were expected to look like flawless Barbie dolls while remaining chaste. They lacked healthy models of sexual desire and agency to inform their choices. A 1997 study, "On Becoming an Object," found that the girls most likely to commit violence against other girls "did not have any sense of themselves or other girls as having their own legitimate sexual desires or being valued." Because girls believed that their sexual value was tethered to both male satisfaction and "idealized standards of femininity," girls attacked other girls who threatened them and their relationships with men. Their paths to power and self-actualization were so pockmarked, it's little wonder some girls turned to violence. The culture so en-amored of the catfight had driven girls to it.

As objectification theory would explain, girls were overtly sexualized and valued only for their sexual appeal, while simul-taneously being denied the tools to understand and command their own authentic sexuality. Agency, pleasure, and safety were increasingly neglected aspects of their development. Still, girls' sexuality caused tremendous fear. There was the risk that if left unchecked it could turn predatory. A few prominent stories of young female sexuality gone horribly wrong quickly became cautionary tales. Amy Fisher was this nightmare come to life.

CASTRATION ANXIETY

In June of 1993, a Virginia manicurist named Lorena Bobbitt also made headlines for violence. The nature of Lorena's crime made her infamous—she severed her husband's penis with a kitchen knife while he slept, then flung it from a moving car into a field, where it was later retrieved and then reattached to its owner. Bobbitt's act realized men's deepest fears about women, at least according to Freud. South of the Mason-Dixon Line, the father of psychoanalysis's castration anxiety was brought to life.

This dramatic and weird crime story was covered far and wide. Reactions riffed off of the darkly comedic element—David Letterman called her "my girlfriend Lorena" for cringe laughs while radio stations looped the song "Re-Attach My Member," which took the tune of the Rolling Stones' "Let's Spend the Night Together." A radio station covering the subsequent trial gave away cocktail wienies and Slice soda. *Newsweek* named it the "cut heard round the world."

Bobbitt's violent act quickly became a grotesque metaphor for the state of relations between the sexes in the 90s. A psychologist told *20/20* that the act frightened men. "A lot of men who are having a rough time in their relationship with the wife are much more nervous and anxious right now," he said. Critic and professor Camille Paglia compared it to the Boston Tea Party and deemed it "a wake-up call" for men. Radio talk shows wondered whether the incident was isolated, or if "violence between men and women has now reached an unimaginable depth." These characterizations prioritized male victimhood, and didn't even begin to tell the whole story.

It soon became clear that the Bobbitts' marriage was rife with domestic abuse. Lorena claimed that her husband, John Wayne Bobbitt, had repeatedly physically and sexually assaulted her, including raping her that very evening, and that she acted in self-defense. Her employer and landlord later confirmed spotting bruises on her body. County police had answered domestic violence calls at the Bobbitt home half a dozen times prior to the incident, and once arrested John for striking his wife in the face. Lorena alleged that she had suffered five years of marital rape.

News outlets did bring forth highlights from this history. But the domestic violence angle hardly got the attention it deserved. "I just remember feeling that there wasn't any traction on the domestic violence part of it," Kim Gandy, then executive vice president of NOW, told the *Huffington Post* in 2016. "Domestic violence organizations tried to have the conversation, women's organizations tried to have the conversation, but the media wasn't having any of it."

Psychologist Elizabeth K. Carll distills the problem with the media's sensationalistic coverage in her 2003 report "Violence and Women: News Coverage of Victims and Perpetrators." Because the mainstream narrative inspired widespread jokes about castration "becoming a worrisome social trend" and caused men to fear "that other women might resort to this kind of sadistic violence," the issue of violence against women was ignored in favor of empty fears about men becoming prey. The problem was that violence against women was "not viewed as a major social problem" at the time, Carll wrote. It was more gripping to imagine how women could be villains than to see how they were victims.

Instead of shining a light on partner abuse, the public and press villainized Lorena Bobbitt for committing "revenge." *Newsweek* determined that the slice was the "handiwork of a bedroom vigilante," and said, "You do not have the right to kill or maim someone you claim assaulted you an hour ago. That's not self-defense. That's revenge." She became "a symbol of female rage." ABC's Hugh Downs called the crime "vengeance of a high-voltage nature," unconvinced that Lorena was acting in self-defense, because the act was mutilation and not murder. "If . . . she had killed him instead of doing what she did, you could think in terms of a self-defense move," Downs said. "This was not self . . . it doesn't seem to . . . I mean, it could have been self-defense. She just made an embittered enemy."

Lorena didn't act like the textbook abuse victim the press and public shamed. ABC news reporter Tom Jarriel presented the other "options" Lorena had as if they were all equal and easy to choose—asking a neighbor for help, obtaining a restraining order, getting help from a friend of her husband's who had been sleeping on their couch—but said instead "she chose the knife." One victims' advocate who initially defended Bobbitt later changed her mind. "Her abuse of him was so barbaric that the fact that she was allegedly abused is hardly an issue," she told the *Los Angeles Times.* Bobbitt should only symbolize a "sick marriage between two angry people," the paper continued, lest she "compromise the legitimacy" of other abused women who "strike back in self-defense." Many Virginians were still unsure whether spousal rape was even a crime. In fact, Bobbitt's cut took place the same year that the last of the fifty states—Oklahoma and North Carolina—criminalized the act.

Bobbitt had told police her husband was selfish in bed and never gave her orgasms, contributing to the narrative that she was either an erotomaniac or sexually frigid. A television series celebrating *National Lampoon*'s twenty-fifth anniversary satirized the incident in an episode called "He Never Gave Me Orgasm: The Lenora Babbitt Story," about a "crazy, sexually frigid hysteric." A popular joke at the time was "How does Lorena feel after sex? She gets a little snippy."

The villainization of Bobbitt continued into her January 1994 trial where she was charged with malicious wounding. Attorneys typically present their opponents in an unfavorable light, and John's attorney, Gregory Murphy, chose to cast Lorena as a jealous vixen. Murphy claimed that his client was seeing other women, and that Lorena was "vindictive and said: 'If I can't have John, no one can,'" wrote the *Los Angeles Times*. He even read Lorena's statement about her husband denying her orgasms on television, as if that were further proof of her erotomania and guilt. Lorena's attorney didn't seem to help matters when he dismissed what his client had said as a problem with a "language barrier" (she was born in Ecuador, raised in Venezuela) and the fact that she was "hysterical at the time."

Because of the violent act she had committed, her claims of rape and abuse were drowned out or their veracity was questioned. Jarriel presses Lorena on her rape claim, asking four different ways if her husband clearly understood that she wasn't consenting to sex. "How emphatic were you when you said no sex?" Jarriel asks. "He never slowed down and he never agreed to stop and he knew, he knew very clearly, that you were fighting him and you were resisting him?"

A judge acquitted Lorena after she pleaded temporary insanity, and ordered her admission to a psychiatric hospital for forty-five days. But feminists who tried to explain that Bobbitt was a victim were attacked—they made her an "instant feminist pin-up girl," lamented a column in the *Baltimore Sun*. A trial spectator called the Bobbitt knifing "every woman's fantasy." The media couldn't separate rape and sex abuse from the comedy and tragedy of a missing penis. Writing in *Mother Jones* about the attack, Katherine Dunn summed up the media's characterizations of Lorena in this way: "The female criminal violates two laws—the legal and cultural stricture against crime and the equally profound taboo against violent females." Both Bobbitt and Fisher breached the barrier of womanhood in this way, and thus were denied their victimhood. They were savaged for their crimes because the ultimate betrayal of femininity is violence, and they had both crossed that line.

PATTERN OF ABUSE

The blame lobbed at Nicole Brown after she was murdered was typical for domestic violence victims in the 90s. Women who suffered domestic violence were faulted for not walking away from powerful abusers, especially when there were children involved. Domestic violence was widely believed to be a personal matter to be dealt with in the home, rather than what it actually is—a crime perpetrated (universally) to oppress women. The tide has since turned against star athletes accused of spousal and partner abuse. But it was in the 90s, during the O. J. Simpson trial, that the country was forced for the first time to confront the epidemic of domestic violence. Sadly, Brown had to die and be blamed for it first.

Eighteen-year-old waitress Nicole Brown met thirty-year-old, world-famous, multimillionaire O. J. Simpson at a Rodeo Drive nightclub, the Daisy. They began dating before his first marriage ended. The couple was living together by the time Brown was nineteen. They married in 1985 and filed for divorce in 1992, the Year of the Woman. Some accounts describe Brown as a devoted mother to her two children with Simpson. Others framed her as a party girl who "zipped around Brentwood in a white Ferrari" (or a Porsche, or a Mercedes) and banked a hefty divorce settlement and monthly child support. Friends called her generous—she frequently picked up checks and gave babysitters hundred-dollar bills to take her kids to dinner without requesting the change. They also recalled her bluntness. She would criticize a friend's bad haircut, or call Simpson an "asshole" if he was being obnoxious in public. Others thought her to be unapproachable. Close friends, like TV personality Kris Jenner, called her shy. "Nicole wanted her space and needed to keep her distance because she was going through a lot during those years," Jenner told journalist Sheila Weller, who wrote the book *Raging Heart* about the Simpson marriage.

Brown's story and character would ultimately be defined by others when she became a world-famous battered and then murdered wife. There were reams of evidence of the abuse she suffered at Simpson's hands. Much of it was made public, some of it even prior to the murders. With only a quick Google search, you can hear chilling audio of Brown's beatings and frantic emergency calls. She sounds terrified and also exhausted as she begs yet another 911 operator to dispatch yet another cop. "He's going to beat the shit out of me," she says in one such call from 1993. "He's going to kill me! O. J.'s going to kill me."

Before Simpson's fame fueled the celebrity spectacle of his murder trial, celebrity worship, particularly by law enforcement, excused Simpson of an established pattern of assault and torture of his ex-wife. Simpson pleaded no contest to spousal abuse in 1989. The Los Angeles district attorney called the police handling of the case "a terrible joke." Police arrived at Brown's home to the sight of her leaping from the bushes covered in blood. Not only was Simpson not initially arrested, he fled the scene in his car. Officers didn't chase his vehicle that time. Police reports logged Brown's screams, Simpson's threats, and the bedlam at the couple's home. But some officers chatted up the former Heisman Trophy winner and courted his autograph. Gutless policing and a pushover judge allowed Simpson to dodge jail time. He breezed into a light sentence of phone calls with a counselor.

Not only did he remain a spokesman for the Hertz car-rental company after his conviction, but he was anointed an NBC News sportscaster later that same year. "We regard it as a private matter to be treated as such between O. J.'s wife and the courts," Hertz's executive vice president said publicly. Simpson defended his conviction at a press conference celebrating this plum gig: "It was really a bum rap. We had a fight, that's all," he assured his new employers and the media. Prominent sports agents speculated that Simpson never would have been hired had the "rap" been a drug charge. But domestic violence was, apparently, forgivable.

The injuries Nicole Brown suffered were many and multifarious. Photographs documented countless cuts and bruises. Police reports recalled her bloodied teeth, bald scalp where her hair had been torn out, and handprints on her neck from strangulation. Brown's journals detail grisly acts of verbal, physical, and psychological abuse, in both private and public. A sushi bar manager recalled that Simpson once repeatedly shouted, "Fuck you, bitch!" to Brown at his restaurant. Simpson stalked her, controlled her, threw her from a moving vehicle, beat her during sex, and threatened her life.

These incidents, taken together, scream of Simpson's guilt. But the prevailing sentiment about domestic violence in the 90s—that it was a private affair to be dealt with at home, not in public, and certainly not by law enforcement—made the case against O. J. less than bulletproof.

A CRIME OF PASSION

Marcia Clark has always called the Simpson-Brown relationship a classic case of domestic violence. But to many, Brown didn't appear to be the pitiful victim. Even friends claimed that Brown wasn't innocent in these disputes, calling her "feisty" and "not a doormat." Weller, the journalist who spoke with some eighty friends, colleagues, and associates of Simpson and Brown for her book about the couple, later said that five people told her independently that Brown "knew how to push O. J.'s buttons." Weller and others in Simpson's inner circle believe that the story wasn't a "classic" case of domestic violence murder, but rather "a crime of passion." And while relationships certainly contain countless shades of gray, the fact is that Brown ended up dead.

"People looked and said, 'Let them handle it in the family. It's not a law enforcement issue. It's not a crime,'" Clark told me. Victims felt the same way. The most common reasons they gave for not reporting abuse to police were the belief that it was private and personal, the fear of retaliation, and incredulity that law enforcement could help, according to a 1998 report on intimate violence by the US Department of Justice. While the prosecution established a pattern of Simpson's abuse in their arguments and offered evidence and witnesses to support this, it wasn't the thrust of their case. Clark says this was because jurors weren't ready to hear a domestic violence argument, especially not when the abuser in question was a beloved sports and movie star.

Legal analysts blamed Clark for not hitting the domestic violence angle harder and failing to convince the jury that Simpson was "a wife beater turned killer." Clark had an "'emotional resistance' to the issue, apparently related to memories of being raped as a teen-ager and having shoving matches with her possessive first husband," according to a review of Clark's memoir. The piece was headlined "The Wrath of Clark," and suggested that her own victim identity inhibited the central argument in her case. It was like double victim-blaming.

Other legal commentators agreed with Clark that a jury would not naturally sympathize with a domestic violence argument. UCLA law professor Peter Arenella said on ABC News that it would be "very, very risky" for the prosecution to "attempt to portray this as a traditional domestic violence case culminating in murder, where the predominant theme is control," because it could "open the door for the defense to show that this woman was not to be infantized [sic]." In other words, arguing that Brown was a domestic violence victim could backfire because plenty of her friends had said that she goaded Simpson. "Back in the day it was difficult to get jurors to believe that she didn't have it coming, she didn't ask for it. It wasn't just a fluke," Clark said. Journalists covering the trial agreed with Clark. "The domestic violence piece was definitely present," former ABC News Los Angeles bureau chief Kathy O'Hearn recalled, "but was the culture as evolved and conscious of it? Probably not."

If Brown or any other woman was being mistreated, why didn't she just leave? How could she stay with an abuser and not protect her kids? Victims' advocates say these are common questions and assumptions that encourage victim-blaming. Clark knew this knee-jerk response to domestic violence all too well. "As of 1994, the prosecutors knew in general that victim-blaming was wrong. I don't think the public was all the way on board back then," she said.

It wasn't just defense teams and media commentators who shaped victim-blaming rhetoric, like the kind that applied to Brown, Lisa "Left Eye" Lopes, and Lorena Bobbitt. There was a broader problem with terminology in the courts and on the streets. Definitionally, at the time, domestic abuse was misconstrued. In the 1970s, psychologist Lenore Walker codified the term "battered woman syndrome" to characterize the "learned helplessness" victims in abusive relationships succumb to in order to survive. This pathology proved useful in early 90s court cases—BWS was often cited to help convict batterers and to exonerate women who killed their abusers in self-defense. In 1991, governors in Maryland and Ohio reduced sentences for women jailed for assaulting or killing partners because they claimed to be victims of battered woman syndrome.

BWS was said to translate the victim experience for society, explaining victims' inability to change their circumstances once trapped in cycles of abuse. It also buffered against victim-blaming. But the BWS theory—which was inspired by a study of dogs that didn't leave punishing environments when given the choice—was later disproved as outdated, unspecific, and stigmatizing to apply to abuse victims, whose experiences weren't all analogous. It wasn't just refuted; experts called it "misleading and potentially harmful" because it created a victim stereotype that not all victims met. BWS painted all domestic violence victims with one brush—the helpless, injured, pathetic woman. If a woman was spirited, like Brown or Bobbitt, she didn't fit the BWS victim stereotype, and her victimhood could be undercut or denied. For instance, when Lorena Bobbitt's lawyer claimed that she suffered from BWS, John Bobbitt's lawyer was able to shred that argument simply by recounting Lorena's crime. Defense attorney Gregory Murphy told *20/20*, "To portray [John] as this person who is very physically abusive based on her word and his denial and to say that she's this poor person that can't do anything, um, is ironic in light of the fact that she's mutilating him."

The Simpson defense team recruited Lenore Walker as a trial witness, to strengthen their claim that Brown was no victim. But Clark correctly avoided a prosecution strategy that focused on Simpson's history of domestic violence. Thus, Walker was never called to the stand. Meanwhile, Simpson's lawyers weren't afraid to discuss his record of domestic violence because they knew they could convince the jury that it wasn't related to the murders. They even renamed it to suit their purposes. Simpson had been involved in "domestic discord," Johnnie Cochran repeated, suggesting mere arguments, maybe pushes or shoves. But they loved each other very much. That was the message. "See, we've been saying for months that the—the case regarding domestic discord was—it should not ever be part of this case. And so finally they accepted that," said Cochran after the prosecution decided to forgo calling some domestic violence witnesses to the stand. Civil rights attorney and trial commentator Leo Terrell denied that batterings were connected to murder. "That's faulty logic," he told an interviewer.

But the logic was not only sound, it was supported by data. A 2003 National Institute of Justice report cited research that in 70 to 80 percent of partner homicides, the man physically abused the woman before the murder, no matter which partner was killed.

BREAK THE SILENCE

Awareness of violence against women and momentum to end it were picking up in the early 90s. In 1993, the United Nations General Assembly for the first time named, defined, and vowed to eradicate domestic violence. It adopted the Declaration on the Elimination of Violence Against Women, which clarified that the abuse of women and girls, whether physical, sexual, or psychological, was not a personal or family matter. It was violence, it was criminal, and perpetrators should be arrested and prosecuted. Each act of domestic violence, whether it was a shove or worse, preyed on women and girls because of their "subordinate status in society," according to a report by a gender-advocacy organization partnering with Johns Hopkins University.

This recognition and expanded definition by the world's most prominent global institution marked an important step. It acknowledged that victimizing girls and women because of their gender was a means of exerting power over them, and it wouldn't be tolerated. "Many cultures have beliefs, norms, and social institutions that legitimize and therefore perpetuate violence against women," the report explained.

Two subsequent global convenings on women's rights in the mid-90s made domestic abuse a central theme. At both the International Conference on Population and Development in Cairo in 1994 and the Fourth World Conference on Women in Beijing in 1995, ending gender violence was framed as a matter of basic equality. At the latter conference, Hillary Clinton famously declared, "Human rights are women's rights, and women's rights are human rights—once and for all," to a raucous crowd.

In September 1994, the year that Simpson was arrested and charged with murder, President Bill Clinton signed the Violence Against Women Act into law as part of his signature crime bill. VAWA for the first time made domestic abuse a federal crime. It funded prevention and victims' services, created special domestic abuse units within law enforcement, and designed trainings for police. It also established a new wing of the US Justice Department to enforce the new law. The bill was cheered as watershed, and encouraged a paradigm shift in how law enforcement and victims' services agencies addressed domestic violence.

After the Simpson verdict, domestic violence awareness peaked. Women's shelters reported greater intakes. Crisis hotlines were flooded with calls and needed more funds to handle the additional volume. California, which in the late 80s and early 90s pioneered legislative remedies to domestic violence, recorded increased arrests. In fact, domestic violence became the number one felony arrest in the state—more than arrests for rape, robbery, and assault with a deadly weapon combined, according to a *San Jose Mercury News* investigation. A Department of Justice report found that intimate partners committed fewer murders in 1995 and 1996 than any other year since 1976. The Simpson trial awakened the nation to a long-simmering scourge, and something was finally being done about it.

But victim stigma didn't disappear so easily. In 1995, Bill Clinton declared October "National Domestic Violence Awareness Month," ironically the very day before Simpson was acquitted. Many pondered whether the verdict would dissuade victims from coming forward and dash hopes that they could get help. Meanwhile, communications campaigns about violence against women seemed to hinder the cause. Posters, brochures, and television ads featured women "often literally cowering in fear and shame," according to a 2012 Avon Foundation report. Critics said such images contributed to the stereotype of "learned helplessness" and victim-blaming. One ad featured a woman crumpled behind a toilet, with the text: "If you're looking for help, you won't find it here. Domestic violence, break the silence."

Toward the end of the decade, the VAWA measures appeared to be having their intended effect. By 1999, the federal government had directed $1 billion to combat domestic abuse, leading the Associated Press to declare, "Five Years after Simpson, War against Domestic Abuse Improves." From the law's passage in 1994 through 2010, some $4 billion of government funds were spent fighting domestic violence. Simpson is name-checked in the first sentence of the 2003 state government report *California's Response to Domestic Violence*, declaring his case as the moment that focus intensified on this "often-hidden form of abuse."

And yet, domestic violence still endangers women far more than random crime. In 2013, women were nearly ten times more likely to be murdered by their husbands or intimate partners than by strangers. The likelihood of a woman being killed by her abusive partner increases fivefold if he can get a gun. Despite the fact that one of the nation's most famous football players was the impetus for widespread domestic violence awareness in the United States, it wasn't until 2014 that the National Football League was forced to change its handling of domestic violence cases. This came after the release of a shocking video in which a Baltimore Ravens running back, Ray Rice, beat his girlfriend, knocked her out, and calmly dragged her limp body out of an elevator. Rice was suspended without pay, then reinstated, and was able to recover financial damages from the league.

Misconceptions about domestic and intimate violence also persist. Victims continue to be discredited to such a large extent that most sexual assaults are never reported to law enforcement. The best that can be said nearly a quarter century after Simpson's acquittal is that blaming victims and ignoring domestic abuse is less tolerated now than it was in the 90s.

11

CATFIGHT

Tonya Harding became one of the most infamous villain bitches of the 90s after she was accused of kneecapping her figure skating rival to keep her from competing in the Olympic Games. But she says millennials don't remember her for this. "People recognize me from the show *World's Dumbest Criminals* and things. That's where they recognize me now," Harding told me. "And then they might Google me because they think I'm so funny and they might come up on skating." She's referring to the truTV series about "brainless bad guys" and "senseless sportsmen" that she appears in as a commentator.

For those who do remember the January 6, 1994, attack on figure skater Nancy Kerrigan, the plotline shifts depending on who you ask. It's not surprising that people misremember the intricacies of the scandal. What is curious is how many people believe that Harding clubbed Kerrigan. In fact, a goon hired by Harding's ex-husband, Jeff Gillooly, struck Kerrigan (with a retractable baton). I had the facts wrong myself when returning to this story for the first time since my youth.

Most recollections of the whacking involve a catfight. When I first Googled "Tonya Harding," the search engine recommended "Who did Tonya hit?" as a frequently searched option. In the countless reenactments that persist today, from downtown cabaret versions, to drag re-creations, to network television sketches, Tonya decks Nancy. In a surreal *SNL* skit, Harding beat up others, too, like John Wayne Bobbitt—he of the severed member. It's not uncommon for people to recall that Harding went to jail. Actually, she struck a plea bargain to avoid jail time but was put on probation, fined, and banned from figure skating for life. But the version of the story that has become stuck in our collective memory is that Harding beat up Kerrigan. In reality, Harding was not personally involved in the assault on Kerrigan. Rather, she admitted to knowing about the incident after the fact and failing to tell investigators.

Most people think Harding kneecapped her rival Kerrigan because she was jealous of her. It was easy to believe when people described Nancy with words like "elegant" and "Kennedy," while Tonya was "hard-bitten" with "rough edges." Female figure skaters have been called the Barbies of sports, so it's no wonder that the promise of a doll brawl leading up to the 1994 Olympic Games in Lillehammer, Norway, was irresistible, even if you knew diddly about figure skating. Skating was simply the backdrop to our favorite kind of fight: girl-on-girl.

The catfight stereotype is so intoxicating and timeless that it inspired the mistaken remembrance that Harding hit Kerrigan. Filmmaker Nanette Burstein spent eight hours interviewing Harding for her *Price of Gold* documentary, which revisits the biases at work around the attack. "Even though the deed was done and organized by men, there was this overriding question of Tonya's complicity," she told me. "I think this was distilled down to Tonya doing the deed herself. That's what we remember from the messaging in the press explosion: the ugly duckling taking revenge on the lovely ice queen."

The pull of the catfight was powerful enough, in a sense, to rewrite history. The events that actually transpired before and after the 1994 Winter Olympics have been all but lost to the erroneous mythology. In the process, both women were villainized, bitchified, and neither was left unscathed.

MUSIC BOX FIGURINE COME TO LIFE

Nancy Kerrigan epitomized the ice princess fantasy of ladies' figure skating. Judges and fans praised her body, grace, and skating style for exemplifying the sport's baked-in feminine ideals. Girlish skaters like Kerrigan were usually young and single. During competitions, cameras often cut to shots of them cuddled with their parents, feeding the fantasy that they were perennial children. One commenter described Kerrigan as a "music box figurine come to life."

Despite Kerrigan's looks, this image didn't come naturally to her. In fact, it had taken her years to cultivate. Growing up in Stoneham, Massachusetts, the onetime "tomboy" had wanted to play ice hockey, not figure skate. But her mother redirected her from roughhousing with her brothers to the more ladylike sport, and her father worked two jobs to fund the hobby.

Kerrigan eventually grew her hair long and developed an effortless-looking skating style inspired by ballet. She perfected and popularized the elegant poses held for seconds while speed-ing across the ice that figure skating is known for. Her coaches choreographed her programs to display her lean, graceful frame. Kerrigan's signature moves rendered her frozen, motionless—an "ice sculpture" meant to be "admired from all sides"—rather than a passionate athlete with a beating heart. Her classical costumes and musical selections were both influenced by and designed to win over skating's rigid judges. Before the attack, the *New York Times* called her the best female figure skater in the country.

Kerrigan's family and background were thoroughly work-ing class, but she benefitted from looking like she came from wealth. She had "a very good patina to her," said *Boston Globe* journalist John Powers in a documentary about the incident. Skating's judges, writers, and commentators seemed to obsess over her "erect carriage" and the "long, slim lines of her body," which "suggest finishing school polish and a kind of haughty dis-tance." Skating in designer costumes furthered the appearance of an heiress. "Doesn't she look like a little angel? Very feminine, very ladylike style," enthused commentator Peggy Fleming at the 1992 US Nationals ladies' free skate. That same year, Kerri-gan won a bronze medal at the Winter Olympics in Albertville, France, and a silver medal at the World Championships.

Skating authorities wanted a princess, and Nancy Kerrigan delivered. Endorsement deals piled up with brands like Reebok and Campbell's Soup. Gone was the tomboy and the cute girl-next-door. The Kerrigan selling shoes and chicken noodles was "supermodel beautiful," cutting a figure compared to Audrey Hepburn, Jackie O, and Snow White. Naturally, Kerrigan ex-pected to take gold at the 1993 World Championships in Prague. But she gave a lackluster performance, only landing two of six choreographed triple jumps. She cried while exiting the ice and took a dismal fifth place.

"She thought she was going to a coronation, not a competition," Kerrigan's coach Evy Scotvold told *Newsweek*, scolding his charge through the press. Something needed to change ahead of the 1994 Olympics in Lillehammer, which were less than a year away. Kerrigan intensified her training, saw a sports psychologist to help her flush out negative feelings, and practiced smiling more.

HER TICKET OUT OF THE GUTTER

Unlike her rival, Tonya Harding didn't need convincing to take up figure skating, despite never fitting the sport's feminine mold. "I stepped out on the ice and loved it the very first time," she told me. "Since that time I was growing up and doing many things that some of the other kids weren't doing." In her biography *The Tonya Tapes*, Harding recalls physical, verbal, and sexual abuse as early themes in her life. She remembers that her mother slapped, kicked, and even "beat her with a hairbrush" in the bathroom at a competition. Harding practiced at a public rink in a local mall, Clackamas Town Center, outside of Portland, Oregon. When she was a kid, her coach paid a competitor five dollars to insult Harding before she skated her program because anger was her greatest motivator. Her "muscular arms and chunky thighs" would earn her the derogatory label "athletic."

Harding's hobbies included drag racing, fishing, and smoking, despite her suffering from asthma. Even her efforts to breathe were unattractive: "Sucking on an asthma inhaler before she took the ice scarcely improved her image," *Rolling Stone* said. Harding broke the fourth wall of ice skating—the space separating performer from audience—by showing her competitive streak in public and at press conferences. She spited the judges with her wardrobe choices, choreography, and musical selections. While most skaters set their programs to demure and tinkling classics, Harding chose ZZ Top and the theme from *Jurassic Park.* "She didn't play by the rules," Burstein said. Needless to say, Harding didn't land the lucrative endorsement deals Kerrigan had.

During competitions, judges wielded tremendous power to jettison skaters they didn't like through the incredibly subjective category of artistic impression. This comprised half of a skater's program score. While most skaters incorporated judges' suggestions and critiques into their future routines, Harding sassed them. In one competition, Harding wore a homemade costume. "One of the judges came up to me afterward and said, 'If you ever wear anything like that again at a US Championship you will never do another,'" she recalled in Burstein's documentary. "I said, 'Well, if you can come up with $5,000 for a costume for me, then I won't have to make it. But until then stay out of my face.'"

Harding rankled the skating world because she refused to conform to the "illusion of decorous femininity" the sport demanded. Skaters told Burstein that Harding's odd, criticized choreography would be considered beautiful today. Skating to rock 'n' roll music isn't outlandish anymore. Now, athletes choose all manner of song medleys, and can even skate to music with lyrics. But back then, this was verboten.

Harding's coach, Diane Rawlinson, a former Ice Capades star who collected wine and "looked like she stepped out of a Ralph Lauren catalogue," counteracted her brashness. Harding's friend and fellow skater Sandra Luckow told me that when she suggested involving child protective services after seeing Harding's mother beat her, Rawlinson declined, knowing that if Harding were to be taken from her home it would kill her skating career. "Skating for Tonya is her ticket out of the gutter. She would have nothing in her life if it wasn't for her skating," Rawlinson told Luckow.

WILD THING

In February 1991, Harding dumbfounded judges and the skating world when she became the first American woman skater to land a triple axel jump in competition. The axel is the only jump requiring a skater to take off while moving forward rather than backward, and the triple is actually composed of three and a half revolutions in the air. It was also regarded as the most difficult jump in the sport, and thought to be the province of male skaters. Going into the National Championship competition, Harding haters had plenty to jeer at, like her strange blend of program songs, including "Send in the Clowns," "Wild Thing," and the theme from the Batman movie, and a costume "the color of Crest toothpaste." But when Harding stuck that jump, she took the national title and became impossible to ignore. The skating world was aghast. "There was no question that Tonya Harding was not the image that figure skating wanted," said Powers.

Witnesses praised the jump while assailing the skater. The *New York Times* said it was "stunning in its athleticism and historic success," while calling Tonya "reckless" for taking such a risk, even though it made history and earned her the title. She received a perfect technical merit score from one judge, a figure not given to a woman skater in nearly two decades. Her so-called recklessness paid off—Tonya was a favorite cruising into the World Championships that year. "If Harding skates a clean program and includes her big jump, she wins, period," declared the *Oregonian*. "Unlike most skaters she almost totally controls her own destiny." Brash, nervy, and full of promise, Tonya Harding was a phenom like figure skating had never seen. Many believed she was unstoppable.

But after 1991, she never did repeat the triple axel successfully in competition. When that athletic achievement was gone, detractors ridiculed what remained: her womanhood. "Her incompetence as a woman . . . marked her as a deviant," wrote Abigail Feder in an essay about ladies figure skating's "overdetermined femininity," citing Harding's masculine hobbies, muscular body, and a reported traffic dispute during which she allegedly picked up a baseball bat. Underscoring Harding's failure as a woman was her inability to keep a man. First, reports of her marriage, at nineteen, raised eyebrows, as skaters her age and caliber were typically single and still lived with their parents. Kerrigan fit both virginal criteria, while Harding "brought impurity to the sport—she brought her husband," wrote critic Laura Jacobs. After Harding took the national title, news outlets asked her how she managed to both skate and tend to her marriage. They also criticized her appearance. "Without those jumps she wasn't much to look at," reporters recalled. Although Harding was a top competitor, failing to be the right kind of woman impacted her financially—"it reduced her value as a television entertainer and commercial spokeswoman long before she was connected to the Kerrigan attack," Feder observed.

Some found Harding's lack of polish refreshing. Journalists described Harding's "rough edges" as if they were a charming accoutrement. To some, her eccentricities "made her intriguing" and proved her bright future. "Not Your Average Ice Queen" topped a punchy 1992 *Sports Illustrated* profile celebrating her quirks. She barely missed medaling at the 1992 Olympic Games and the 1993 Nationals, placing fourth both times. By the time she headed to the Nationals in 1994, she was hungry and poised for the podium.

WHY ME?

On January 6, 1994, before the United States Figure Skating Championships in Detroit, a hit man clubbed Nancy Kerrigan's leg and fled the scene, crashing through a glass door. The evening news broadcast footage of Kerrigan crumpled on the ground and shrieking "Why?" and "Help me!" in the moments after the attack. One clip cuts to B-roll of her father whisking her to safety, swaddling her like a newborn in his "powerful welder's arms." There was requisite pity for Kerrigan, but it wasn't long before the scoffing began. The "Why?" heard round the world was spliced, replayed, mocked, and parodied. It morphed into "Why me?" on *Newsweek*'s cover—conveniently editing out the "help" that Kerrigan was calling for—over a close-up of Kerrigan's anguished face. Glimpsing raw emotion was uncharacteristic in the hermetic world of ladies' figure skating in the early 90s, which rewarded the projection of virginal girlhood and skaters' icy veneers. A self-described father of three sporty daughters asked, "Has anyone noticed what a crybaby she is?" in the *St. Louis Post-Dispatch*. He called her an "embarrassing sight" and suggested she was "mentally soft."

Skating purists and journalists lamented that the assault deflowered the princess sport. Kerrigan's attacker was as guilty of jeopardizing her prettiness as he was of injuring the skater. Her beauty was "distorted as she watched a life's work perhaps ruined," lamented *Newsweek*. "We don't want to look at Kerrigan and be reminded of how ugly the world can be," complained the *Denver Post*. "When the attacker struck Kerrigan, it nauseated us. We felt defiled, too." The perpetrator "took a crowbar to porcelain legs," "defaced a beautiful symbol," and "raped a sports myth by beating Kerrigan." Sensing that the sport itself was under assault, fans leapt to defend skating rather than the athlete. "You want to believe that the beauty will ultimately win," sighed a columnist in the *Atlanta Journal-Constitution*.

A violent act had not only punctured the innocent facade of figure skating—it also made ice princess Kerrigan a commoner and a crime victim. Many were glad to see the fairy-tale fantasy quashed, and were quick to kick the pedestal out from under her. The *Denver Post* wrote, "The Olympic princess became just another crime statistic," and claimed that Kerrigan was guilty of "mistakenly believing she was immune to hate." Some doubted the seriousness of her injury and questioned whether her bruised knee was career-threatening. Others felt she wasn't beaten up enough to claim real abuse. "The skater's physical pain wasn't so severe as that suffered by millions of lesser-known women battered in America every year," reminded one commenter. Gillooly told the FBI that Kerrigan's attacker tried to send cops on the trail of a demented fan with a fake ransom note that read "All skating whores will die" in letters clipped from magazines. Gillooly said that the hit man failed to drop the note but did manage to shout, "I just spent 29 hours on a bus for you, bitch," referring to how far he had traveled to maim her.

Kerrigan's delicate image and her sport's decorum stood in sharp contrast to the violence she'd suffered, hence the eagerness with which followers of the story gawked at and bitchified her. They called her entitled and questioned whether she really was a victim at all. The beating also catapulted her to a kind of fame she'd never known, the kind of notoriety being an Olympian previously hadn't brought. People who didn't live and breathe figure skating suddenly knew her name. Some accused Kerrigan of profiting from the hardship. Suffering was worth the celebrity it begot. An "almost-anonymous practitioner of a marginal sport" was now "one of the most famous athletes in the world," according to the *San Francisco Chronicle*. The subtext: quit whining. It wasn't until much later that Kerrigan spoke of the psychological damage of the attack, the nightmares, her rehabilitation under the glare of stalking cameras, and the mounting self-doubt preceding the 1994 Olympics, where she was expected to not only recover, but to skate the best program of her life and win gold.

Indeed, it wasn't until after Kerrigan had been bludgeoned that the media decided her Olympic gold medal was *predestined*. It was almost as if she had to get clobbered to deserve it. Meanwhile, Harding, who had won a gold medal at the United States Figure Skating Championship in Detroit, wasn't immediately implicated in Kerrigan's takedown. She became linked to the beating a week later when her occasional bodyguard, Shawn Eckhardt, and her ex-husband, Jeff Gillooly, were arrested and charged with criminal conspiracy to commit assault. Blame quickly shifted from the men to Harding, despite the fact that she hadn't been charged. To prove her guilt, detractors cited examples of her failure to be a proper woman, her unsavory character, and the fact that she'd never been welcome in the sport.

Later, Harding revealed that she had known about Eckhardt and Gillooly's role in the attack, but not until afterward. Despite her failure to come forward with that information, she maintained her innocence then, as she does now. Reports revealed that Harding's ex-husband, along with a cabal of wannabe hit men, had brainstormed a more gruesome fate for Kerrigan than what befell her. Someone suggested slicing her Achilles tendon or killing her. They had settled on a career-ending blow to her knee, but hit man Shane Stant whiffed, bruising her leg instead. Kerrigan told reporters she considered herself lucky for that.

TONYAGRAMS

By the middle of January, Harding and Kerrigan were everywhere. The obsession with the spectacle filled the news. *Nightline* anchor Ted Koppel said Harding and Kerrigan were more intensely covered than the fall of the Berlin Wall. Television crews camped out on Kerrigan's lawn to ambush her when she left or returned home. They trailed her to training and physical therapy, begging her to talk. Updates, no matter how small, kept the story alive. Interest in anything tangential to the tale was so great that Kerrigan's hometown paper even interviewed a middle-aged resident also named Nancy Kerrigan who couldn't leave the house without being asked whether she'd been whacked. "In the little way it has affected my life, I just can't imagine what this must be doing to hers," she said.

The skaters and their surrogates appeared on *Sally Jesse Raphael, Inside Edition, A Current Affair,* and many other shows. Connie Chung bagged an exclusive with Harding, and Ted Koppel with Kerrigan. Women were paid to impersonate them.

Tonya Harding look-alike Lynn Harris fell into the gig while living in Boston—"Nancy Country." Harris took skating lessons on her lunch break. Sporting a blonde ponytail and ice skates slung over her shoulder, she was a dead ringer for the embattled skater. She was so frequently stopped on the street that she began to wear a hat and sunglasses so she wouldn't be "recognized." "I had to leave home early so talking to people wouldn't make me late for work," she told me.

After a newspaper article about her Tonyaness, Harris fielded calls from producers, casting agents, and folks wanting her to deliver "Tonyagrams" or camp out on the real Nancy Kerrigan's lawn to "see what happens." Harris won Geraldo Rivera's "infamous celebrity look-alike contest," played Tonya Harding in a musical cabaret show in Greenwich Village, and shocked talk show host Ricki Lake's audience with a surprise faux brawl during an episode feting lesser-known catfights from previous shows. The catfight trope, standardized on daytime television, was useful for understanding that Harding and Kerrigan, already actual competitors in the sports arena, were competing for femininity, too, in the eyes of society.

But femininity can't be won, of course. Author Leora Tanenbaum says catfights are a societal construct. "The dynamic has nothing to do with chromosomes or hormones; it is rooted in a narrow, constricting view of femininity," she told me. Her 2003 book, *Catfight: Rivalries among Women—from Diets to Dating, from the Boardroom to the Delivery Room*, claims the catfight is born of the paradox that society conditions women to compete against one another to achieve peak normative femininity, while warning that competition itself is unfeminine. "We resolve this paradox by competing but pretending not to, and ultimately the indirect aggression bubbles over into what people call a catfight," Tanenbaum said.

GUILTY UNTIL PROVEN INNOCENT

Harding won't discuss the 1994 incident in detail, but she did say, "If anyone would have listened to the media, they would have known that I was asleep" when Kerrigan was struck. *The Price of Gold* hints at a version of the attack story that Harding groupies want to believe—that she didn't know much about it, or anything at all. "I hope that it's true what she's always maintained, that she did not know," says Elizabeth Searle, who wrote the libretto for *Tonya & Nancy: The Rock Opera.* "She was desperate. Coming from a violent background, under the thumb of a violent husband, and this pressure cooker sport. I think the truth often lies in between."

But Harding's trashiness and "failure as a woman" implied her guilt in the Kerrigan attack more than any tangible evidence. Critics demanded that Harding be cut from the 1994 Olympic team. Kerrigan's camp told reporters that it wouldn't be fair for her to compete alongside Harding, even though Harding hadn't been indicted. News stories reminded the public that figure skating had "never warmed to the combative Harding," who was "a little barracuda." And skating officials wanted her head. "If only Harding could be charged, convicted and incarcerated immediately if not sooner, so the games could go on without evil," wrote *Chicago Tribune* writer Bob Verdi, imagining skating officials' views.

Harding's oddball, charming "rough edges" sharpened. She even internalized the characterization of herself, and fed it back to the press that had created it. "Despite my mistakes and my rough edges, I have nothing—I have done nothing to violate the standards of excellence, of sportsmanship, that are expected in an Olympic athlete," she told reporters, pleading her case to compete in the 1994 Games. They wrote that her voice and body were "trembling," "wavering," and "quivering." She was also capitulating to the terminology they had used to define her.

The United States Olympic Committee scheduled a disciplinary hearing to determine whether Harding should compete in the Games, but then canceled it when she threatened them with a $25 million lawsuit. After a vote of confidence from President Bill Clinton, and amid fears that she would wage a legal war if thwarted, Harding was ultimately allowed to compete in the Games. Many were displeased and unwilling to hide their agita, including skating officials. One anonymously told the *Washington Post*, "I can't believe she's actually coming. I can't believe everyone ended up doing nothing."

CAT'S EYES AND CHICLET TEETH

In the lead-up to the February 1994 Olympics in Lillehammer, the details of the skaters' lives that had once humanized them in magazines and ad campaigns were now picked apart and scrutinized. After the attack in Detroit, the press seemed to want the pair to finish each other off. Kerrigan tried to keep it classy and direct media attention to her skating, or away from her entirely. She didn't want to discuss competing against Harding. "I certainly had not gone to Detroit to win a contest against Tonya Harding," she said. Harding, ever the competitor, lamented being denied the chance to beat Kerrigan after the attack. "I worked my butt off for this, and if anybody wanted to beat Nancy, it was me. Who wanted to compete against her the most? It was me," Harding said.

Most striking was how black and white the narrative became. Harding's "dark-hearted sporting she-devil whose claims to innocence have convinced few" was no match for Kerrigan's "snow-white" purity. "It was the supreme princess versus the trailer trash ignoramus," recalled Luckow. "It was so rich in its blacks and whites. Nobody did the gray." A catfight was so widely desired that the women were even described as feline. Harding was "an alleycat" who would "fight and claw." After the attack, Kerrigan became "a different cat," according to her coach Evy Scotvold, with a "peaceful determination" and the "confidence of a gunslinger." Others imagined a slinky, sly Kerrigan with "cat's eyes and Chiclet teeth."

Many girls growing up in the 90s chose sides, and were more likely to favor Kerrigan the ice princess than Harding the trashy villainess. Hilary Bauer was firmly on team Kerrigan as a young ice skater in New Jersey. She wore costumes to match Kerrigan's, and held pictures of her torn from *Sports Illustrated* while watching her compete on TV. "Nancy was the ice princess. She was so beautiful. Tonya was a loser," she told me. Bauer describes the attack as "how I learned the difference between right and wrong." "Honestly, I gravitated toward Nancy because she was so pretty and elegant and I loved her costumes," Jenna Leigh Green, the actress who portrayed Nancy in the rock opera, told me. "As a small child those are the things important to you." I haven't met a woman who followed this story as a girl who identified with Tonya Harding.

BROAD STREAK OF BITCHINESS

As the Lillehammer showdown neared, fans wondered whether the desired catfight would occur. Would the skaters sleep in the same dorm, dine together in the cafeteria, or meet on the rink? New details emerged daily. The number of reporters covering the Games, particularly from American news outlets, increased dramatically—and not just skating journalists, but "a media scrum worthier of Charles and Diana's wedding, the Israel-PLO peace treaty or Nelson Mandela's release from jail," declared one newspaper. The pair shared practice ice, which drew some four thousand fans to watch them warm up. Kerrigan arrived wearing the same dress she was attacked in. Gauntlet thrown. Mark Lund, founder of *International Figure Skating* magazine, said, "Peace in the Middle East, what was that about? This was like watching *Dynasty* in real life."

To the delight of her critics, Harding didn't medal at all. She abruptly left the ice during her Olympic skate because her boot malfunctioned. Judges let her start again. Coming in a dismal eighth was comeuppance. Kerrigan skated the program of her life and was sure she'd won gold. "For me, in my mind and my heart, I did," she told the *Los Angeles Times*. "I was great." Coaches and Kerrigan herself called her routine "flawless." When the gold medal was instead handed to the sixteen-year-old Ukrainian skater Oksana Baiul, many were shocked.

To the public and the press, Kerrigan didn't win silver in the 1994 Olympics. She lost gold. Fans immediately turned on her. Kerrigan hadn't even cleared off the Olympic ice before her princess varnish was blown off. Her "broad streak of bitchiness" first emerged during a delay preceding the medal ceremony, explained *Rolling Stone*. Officials were searching for a recording of the Ukrainian national anthem to play, but Kerrigan thought that Baiul was causing the delay by applying more makeup. "Oh, come on. She's going to get out here and cry again. What's the difference?" Kerrigan said. Little did she know, her mic was on and fans heard the diss. It was a reminder that beneath skating's sparkles, it was still a game of winners and losers, like any other sport.

Some in the American skating community believe that international judges may have denied Kerrigan the gold medal because of the attack, even though she was the victim. Tackiness and Americentrism tarnished the Games, and Kerrigan was to blame, alongside her attackers, for the luridness. Kerrigan added more boorishness with her dig at Baiul, and later by not marching in the Olympics closing ceremonies, and giving "curt answers at press conferences." The media would punish her for being a sore loser.

The girl who had mugged for Campbell's Soup, drank milk out of a champagne flute, and wore designer skating costumes that resembled wedding dresses was now starring in her own horror movie, and those who got off on schadenfreude were eager to see her hacked to bits. When asked to comment on Harding's program, Kerrigan sniped at her inability to skate at her set time due to a broken lace. "They're bending lots of rules, I guess," Kerrigan said of the judges. Once she knew the press had soured on her again, Kerrigan didn't hold back. After telling a teammate that she "sucked" during a skating television special back in Providence, Rhode Island, she turned to the camera and said, "You probably just loved that." Princess of rectitude no more.

A trip to Disney World to kick off her multimillion-dollar partnership with the Magic Kingdom went afoul when Kerrigan was again caught on a hot mic complaining, "This is so corny, this is so dumb," alongside Mickey Mouse and Goofy. An executive shared fears that working with Nancy might become "a nightmare," and wondered if Disney should have chosen a less controversial skater instead. "Overnight, she risked becoming the Shannen Doherty of the skating world," according to the *Washington Post*, referring to the *90210* star whose rumored bitchiness launched a club dedicated to hating her. "Is Nancy a bitch?" the paper asked, citing her "cat's eyes" and "Chiclet teeth" as possible proof. Now that Kerrigan had become the "defrocked Cinderella," haters piled on.

The news media created a spectacle of Kerrigan's suffering and then blamed her for causing such a stir. *Sports Illustrated* admitted she'd "been burned," but accused her of holding the match. "There is no doubt that many of the burns have been self-inflicted: words that shouldn't have been spoken, attitudes that shouldn't have been taken, a defensiveness, almost an animosity toward the great media dance that surrounds her—and helps hype—her booming ice skating career. By now it's hard to figure out who burned whom." A *Boston Globe* columnist called her "a semi-celebrity who, if she couldn't skate, probably would have been saying, 'That's $11.50, please. Pull up to the window for your burgers and fries.'" The champion was now no better than a fast food chain worker. Nancy was getting Tonya-ed.

The spectacle continued to dominate headlines and nightly news broadcasts. Kerrigan's parents pushed back in the press, which had an infantilizing effect, but they had good points. "What if she had been a man?" her mother asked in *Sports Il-*

lustrated. "Would there have been any of this? If she had been a hockey player, she could have been in a fight, and no one would have said anything after the game. Do you think any of this would have happened to a man?" Her father added, "What if she had won by a tenth of a point instead of lost by a tenth of a point? There wouldn't have been any of this."

Harding pleaded guilty to hindering the prosecution in March. In June, the United States Figure Skating Association barred her from skating competitions for life and snatched back her national title. In fact, Skategate pandemonium didn't relent until a new tabloid drama emerged to unseat it. That was when Harris's Harding look-alike gigs dried up. "I lost my job to O. J.," Harris said.

"NO ONE EVER WANTED TO LISTEN"

Today, Tonya is getting Nancy-ed. Sort of.

Like religious parables, Shakespeare, and Greek mythology, the folklore of Skategate is being reappropriated, in many cases by millennials who watched the attack on television as children, but now, more than two decades later, romanticize it as adults. Current retellings hinge on redeeming Harding, the 90s villain-ess, but also celebrating her kitsch. Tracy McDowell, the actress who portrayed Harding in the rock opera, dressed as Harding for Halloween. Maybe there were always Harding apologists, touched by the hardship she overcame, but they weren't loud enough to stifle her detractors. Her new champions are drawn to her antihero qualities, as well as her tacky outfits and sassy mouth. "It's amazing she didn't kill anyone," said Matt Harkins, one such self-selected custodian of Harding's story. Today, Kerrigan lovers are harder to find.

Harkins and Viviana Olen are friends in their late twenties; for them Tonya and Nancy are no passing interest. The Tonya Harding Nancy Kerrigan 1994 Museum in Brooklyn is their creation. The permanent collection is an amalgam of photographs, memorabilia, and original artwork. Hawkins and Olen aren't art historians, but rather comedians who became roommates and then creative partners. Museumgoers have included former figure skaters, an Olympic ice dancer, millennials and their parents who are in town, the cast of *Tonya & Nancy: The Rock Opera*, and swarms of journalists. "There's so much you didn't realize," Olen said. "What a parody Tonya became. She is the first American to land the triple axel in competition and this is something that I have never heard before in my entire life and that's insane to me, you know? It was all about her hair and she's just like a thug or something."

The museum contains news clippings from the incident, as well as pins, programs, and trinkets from competition. Harkins and Olen won't feature anything that mocks the skaters or the incident—despite the cornucopia of that kind of thing—and they steer clear of Harding's boxing career and the DIY pornographic film she and Gillooly sold to *Penthouse*. "We do feel a responsibility," Olen says. "We don't want to focus on the parody unless we're talking about how insane it was."

Often museumgoers are women who loved Kerrigan as girls, who closed ranks around her and bad-mouthed Harding. Olen tells about a visitor who mourned after the attack, covering her windows in Kerrigan pictures and refusing to go to school for two weeks. "I remember it was 'Tonya's bad, Nancy's good,'" said twenty-nine-year-old Clara Elser, an actress who debuted the solo show *The Love Song of Tonya Harding* at the Calgary Fringe Festival. The plot is that Tonya Harding is hosting a viewing party during the Olympic figure skating ladies' long program twenty years after she competed, giving her an opportunity to tell her story. Elser aims to treat Harding with sympathy, and notes the many people who erroneously think Harding did the whacking.

Most of the efforts to retell the epic of Nancy and Tonya focus on Harding. She is the central figure of the documentary *The Price of Gold*, perhaps in part because Kerrigan didn't participate, but also for the same reasons filmmakers, playwrights, musicians, and comedians are drawn to Harding's flaws. When I ask Harding what she thinks of the current retellings of the 1994 story—such as the museum and films, including a biopic starring Margot Robbie—that try to tell parts of her tale that were being left out at the time, she interjects, "No, actually nothing was ever left out. No one ever wanted to listen. That's what's hard."

Plenty of people want to bend the story back to its 90s shape. When New York television station PIX11 filmed at the museum, producers asked Olen and Harkins to reenact the hit (they refused). A Fusion newscast used a Harding Barbie doll to strike a Kerrigan one. Even Barack Obama promised he wouldn't pull "a Tonya Harding" and pummel his opponent, though he might want to, while campaigning for president. Stars of *RuPaul's Drag Race* mocked Harding when they adapted Skategate for a comedic miniseries, *Ice Queens*. "Kerrigan is looking strong at practice while Tonya just ate a meat-lover's pizza. How will the saga end?" a reporter asks as Tonya inspects her fingernails. Instead of sticking her "tricky combination," Tonya "fell, farted, and barfed."

Luckow thinks what's preventing Harding from a real comeback, and redemption, is an apology. "Tonya never apologized or admitted to being a part of it. It's not the act itself, it's the denial that gets people really angry," she told me. "If she had nothing to do with it, fine, but by not apologizing, that is what angers people."

In *The Price of Gold*, Harding's emotions range from anger to tears. She is the one who lost her career and her life's work after the incident, but many focus on her complaining. As *Slate* put it, Harding is "still bitter after all these years." In the lead-up to the release of the major motion picture *I, Tonya*, Harding exercised to get in shape, "redoing her hair and looking over her wardrobe," her agent, Mike Rosenberg, told me. He relays that Harding fears interviews because she's been burned so many times. "The only way she can really get work is trading on her infamy," Burstein said. There have been many comeback attempts. But this time could be different.

Harding still skates, sometimes a few days a week. She works with a coach, who tells her to show up on time since her presence motivates young hopefuls to skate harder. Her advice to young women is a dose of tough love. "Believe in yourself and trust yourself. Love yourself. Because the only person that has to love you is yourself." She tells them to pursue their dreams but cautions that achievement is all-consuming. "It's like me trying to do my triples or something. Pardon my language, but if you half-ass something, it's not going to happen. It's either all in or all out."

After the attack, Kerrigan went on to more endorsement deals and continued skating at paid events like Skate America and Halloween on Ice. She is mostly out of public view, except for a recent stint on *Dancing with the Stars* and yearly performances in Nancy Kerrigan's Halloween on Ice show, which tours midsize East Coast and Midwest cities each October. "She didn't really skate that much and just sat in the chair as a princess," remarked one attendee.

Not long ago, I was riding the subway reading a copy of *Women on Ice*, a 1995 collection of feminist essays about Harding and Kerrigan. I noticed that a guy, probably midthirties, was reading over my shoulder. "Oh, I remember that story," he said. "That white trash girl beat up the ice princess. That story was everywhere."

12

THE GIRL POWER MYTH

I n the early 1990s, new research raised red flags about adolescent girls in America. The 1991 study *Shortchanging Girls,
Shortchanging America* asked nearly 3,000 young people—
2,400 girls and 600 boys in grades four to ten—about their
attitudes toward the classroom, gender, self-esteem, and ambition. The result was that, for the first time, researchers directly
linked girls' poor self-esteem to what they learn and how they
are treated in school.

Shortchanging Girls, Shortchanging America found that girls
and boys had similarly high levels of self-esteem, achievement,
and happiness in elementary school. But as they reached middle
and high school, girls reported declines in all these categories.
Boys were nearly twice as likely as girls to argue with teachers when they believed that they were right, and to assert that
they were "pretty good at a lot of things." As girls and boys
grew up, the achievement gap between them widened, and girls'
self-esteem plummeted further. These findings first set off alarm

bells in 1991, when the study was published. Advocates for young women, not to mention their parents, wanted to know *why*. What caused girls' happiness and confidence to nose-dive in the 90s?

Another major work that raised concerns about girlhood was the 1992 book *Meeting at the Crossroads: Women's Psychology and Girls' Development*, which probed the interior lives of adolescent girls. Harvard psychologists Carol Gilligan and Lyn Mikel Brown spent four years interviewing nearly a hundred girls ages seven to eighteen at a Cleveland private school. They hoped to learn "what, on the way to womanhood, does a girl give up?" Quite a lot, it turned out. Gilligan and Brown found that the stuff of growing up—taking risks, acquiring independence, and expressing desire and voice—was diametrically opposed to the qualities the culture most celebrated in girls, such as niceness, selflessness, rule following, and conventional beauty. "The passage out of girlhood is a journey into silence, disconnection, and dissembling, a troubled crossing that our culture has plotted with dead ends and detours," the authors wrote.

These experiences led Gilligan and Brown to overhaul their traditional research model and devise a new one, which they called a Listener's Guide. Interviewing girls over the course of years allowed them to see and hear girls lose their confidence and power, devolving into self-conscious people-pleasers over time. It also helped them compare the changes in girls' speech by "following the pathways of girls' thoughts and feelings, of distinguishing what girls are saying by the way they say it."

Gilligan and Brown came to understand that when a girl switched from speaking in the first person to the second she was often disassociating herself from her own feelings. For instance, one of the girls in the study, Noura, wrestles with whether or not to confront friends who trash another friend behind her back. She knows it's the "right thing" to do, but she fears disownment from the group. "I learned that it's not nice [to talk about people] and I learned what it feels like to be—to be the person that everyone doesn't like," she tells the researchers. When they ask her how she learned that, she switches to the second person. "You learn that by just your feeling, what you feel was right or wrong . . . You learn because you know that you won't like that to be happening to you." Clearly this has happened to her, but she redirects and broadens the experience to off-load its pain.

Most notably, the researchers found that all of the girls in the study contended with the cipher of the perfect girl. They avoided speaking up or challenging the status quo because they believed doing so would deny them the possibility of being perfect. Even though they knew the perfect girl was neither achievable nor real, the idea that they *could* be perfect still tortured them.

SELF-ESTEEM FROM THE OUTSIDE IN

In the years following the publication of *Crossroads* and *Short-changing Girls*, a host of other books and studies continued to investigate both the interior and public lives of adolescent girls. One such work that confirmed the initial research was *Reviving Ophelia: Saving the Selves of Adolescent Girls*, which indicted a 90s culture that "splits adolescent girls into true and false selves." It argued that girls abandon their authentic selves to kowtow to warped cultural demands—thinness, beauty, and sexual availability.

Author and clinical psychologist Mary Pipher treated adolescent girls who cut themselves, acted out, suffered from eating disorders, and reported anxiety and depression, validating a whole range of experiences, including some of my own. I recall the sad blonde girl on the book's cover because of the length of time it sat on my mother's nightstand. I remember thinking that she was threateningly pretty.

Pipher identified numerous mixed messages that girls coming of age in the 90s contended with. "Be beautiful, but beauty is only skin deep. Be sexy, but not sexual. Be honest, but don't hurt anyone's feelings. Be independent, but be nice. Be smart, but not so smart you threaten boys," Pipher wrote. She concluded that America in the 90s was a "girl-destroying place," even more so than earlier decades, like the 60s, which she romanticizes in comparison. "Many girls say they wish they had lived in those times," she writes. "It's much harder to be idealistic and optimistic in the 90s." Pipher's framing resonated. *Reviving Ophelia* spent twenty-six weeks on the *New York Times* bestseller list.

I now see echoes of the many girlhood struggles catalogued by researchers in my own adolescence. I, too, ricocheted between emotions, capitulated to girly urges, and slaved to placate others. Looking back, what's most insidious is that I had little inkling of the boxes that had been drawn around me by society. Sexually available, yet virginal. Polite, and not too smart. Helpful, never harmful. In the 90s, I accepted the shape I was taking and didn't often question why it wasn't more malleable.

But *Reviving Ophelia* was written for parents, specifically mothers, and not for the adolescents of the 90s. Girls themselves turned to ever-popular teen magazines to make sense of their own struggles—magazines that Pipher blamed for girls losing their self-esteem in the first place. As explored in chapter 2, it's no surprise that these magazines exploded in the 90s. An array of glossies dangling phantom perfect girls lined the lowest supermarket shelf and stoked girls' insecurities. While *Short-changing Girls* and *Crossroads* confirmed that girls were losing themselves on the path to adulthood, influential teen magazines only reinforced that girls should pursue unachievable perfection. The desperate desire to "cure" what was plaguing girls spurred seminars, school assemblies, more research, and lectures. The alleged antidote was something called "self-esteem." It was peddled as a shield to gird against the dangers of sex, the lure of drugs, and attendant pressures from peers. The self-esteem panacea underscored an unquestioned binary: girls without it succumbed to these awful things; girls who had it did not. Thus, "self-esteem" became a buzzword, used in government health programs, elementary school curriculums, and homes. It was sold as elixir for the girl crisis.

We were told to "have" self-esteem and, if we didn't have it, to "get" it. But nobody told us precisely how. Teachers and counselors enumerated the risks of not having self-esteem, or losing it, if you happened to believe you were one of the few who had it already. But it felt like something left to chance, something that you could catch—like a ball or a cold. The flip side of the self-esteem cure was that girls were discouraged from admitting that they didn't have any. Whether or not you had self-esteem was a hard thing to know while straddling the line between childhood and adulthood. It seemed to be something adults could know about you before you knew it about yourself.

All this panic and confusion offered a fertile place for America's favorite pastime: capitalism.

SELLING WHAT GIRLS WERE LOSING

Former schoolteacher and textbook author Pleasant Rowland conceived of American Girl dolls after she couldn't find acceptable ones to buy her nieces. "There was nothing that really spoke to a little girl's soul, that nourished her self-esteem, her sense of quality. There wasn't something to treasure intellectually or physically," she told an interviewer. Rowland put up $1 million that she had earned from textbook royalties to start the Pleasant Company in 1986. It would grow into a commercial empire that packaged and sold self-esteem as a consumer product.

Rowland's playmates weren't just dolls to dress, but characters with rich backstories emerging from American history. American Girls were sold in exclusive catalogues for more than eighty dollars each, but their educational books and era-appropriate accessories, costumes, and furniture ran families hundreds more. The company even manufactured real-girl versions of the historical doll clothes. I begged for years for Samantha, a Victorian-era orphan who lives with a wealthy aunt, but also for two versions of her red taffeta party dress trimmed with a lace neckerchief—one for her, the other for me. Samantha was the company's bestselling doll, perhaps because many girls fantasized about a plush life sans parents. In 1992, the company added *American Girl* magazine, a "less fashion and boy-obsessed alternative" to other periodicals on store shelves that promised to "affirm self-esteem, celebrate achievements, and foster creativity in today's girls." It still calls itself the largest magazine for girls ages eight and up with a circulation of over 325,000.

What set these dolls apart, according to their creator, was

the qualities she hoped they would transmit to girls: "confidence, honesty, innocence, and courage." The premise was seductive. Experts agreed that dolls could act like real friends and role models to girl owners. Here was self-esteem the American way—for sale. And it was pricey. Total US doll sales hit $2.7 billion in 1992. The Pleasant Company recorded $150 million in sales in 1993 alone. At the time, American Girl HQ counted 350 employees to run this self-esteem farm and fielded some fifteen thousand calls a day from customers.

The original crop of American Girl dolls was all white, which not so subtly—and rather sinisterly—suggested that only Caucasian girls deserved the self-esteem Rowland was selling. Not coincidentally, white girls made up the majority of the Pleasant Company's customer base. That Barbie creator Mattel had made a black doll as early as 1968 made American Girl's exclusion even more glaring.

The American Girl dolls were marketed as more civilized, educational, and confidence-producing than trashy Barbie, whose dreams were limited to a self-serve ice cream parlor and a convertible. With girlish bodies and identities that weren't sexualized, they were seen as an improvement upon Barbie, who, by the 90s, was attempting to beat her shallow rap. In 1992, the Year of the Woman, Mattel released Presidential Barbie. "Totally Hair" Barbie—launched the same year with "ten and a half inches for styling!"—swiftly outsold the ambitious politician and would become an all-time bestseller. If you can't beat them, of course, you can always acquire them. Mattel bought the Pleasant Company in 1998.

Psychologists pegged Barbie and American Girl as promotional devices for self-esteem that didn't actually deliver it.

"Open an American Girl catalogue, or take a walk through Limited Too, and you'll see stereotypes of girls with very limited choices about who they can be alongside continuous pleas for them to shop, primp, chat, and do things girls are 'supposed to do,'" warned Lyn Mikel Brown and Sharon Lamb in *Packaging Girlhood: Rescuing Our Daughters from Marketers' Schemes*. "In fact, be aware that every time the phrase 'girl power' is used, it means the power to make choices *while shopping*!"

Dolls were hardly to blame for the 90s girl crisis, but it's hard not to see how they underscored girls' belief that self-worth was found in appearance, not achievements. And this lesson was being absorbed by younger and younger girls. Mattel admitted that it marketed its busty dolls—which earned the company $1 billion a year—to girls even younger than seven years old. One report found that three-year-olds "choose Barbie over baby dolls."

In the 1990s, brands like Mattel and American Girl encouraged girls to purchase self-esteem and chase perfection. Magazines and marketers sold girls goods and beauty ideals to make them feel better about themselves. As objectification theory explained, girls were taught to internalize the outside criticism of their bodies, and to make it the primary view of their worth. The atmosphere was ripe for bitchification to thrive.

ACTING LIKE CHILDREN

By the mid-to-late 90s, the Riot Grrrl revolution had all but fizzled, while the troubles of girlhood, identified by researchers, persisted, if not intensified. Young women still struggled with body image, depression, and the specter of the perfect girl. Many

THE GIRL POWER MYTH 279

still wanted to fulminate against the societal construction of girlhood, but they saw how women perceived as angry, manly, and difficult were summarily bitchified, so they opted to rebel by embracing girlhood but rejecting feminism. Exercising the right to be girly seemed like a logical path of self-actualization. Choice, after all, was a feminist plank. Moreover, plenty of girls weren't ready to be the "women" their mothers' generation told them they had to be once they turned eighteen.

This was the backdrop against which Girl Power, a 90s mantra and marketing tool, rose to prominence. Riot Grrrl was never afraid to show its claws, but Girl Power, its successor movement, lacked teeth; it embraced the "girl" part but did away with everything else. In other words, by the late 90s, the Riot Grrrl rebellion had been diluted and commercialized. Ultimately, it ceased to be a rebellion at all, and instead became a full-on shopping spree. Where Riot Grrrl had used anger and disobedience to fight for social justice, Girl Power revered signifiers of girliness, especially those that could be purchased. Girl Power was such an effective marketing tool that girls growing up in the 90s, myself included, had no inkling of its origin as a protest movement. We thought the Spice Girls had invented it. To us, at the time, Girl Power appeared as an opportunity to delay womanhood and gain imagined parity with boys. It was "becoming a woman on a girl's own terms," according to the *Los Angeles Times*. Geri Halliwell, a.k.a. Ginger Spice, crystallized this strategy when she told a reporter that feminism needed weakening to become more palatable. "It's about labeling. For me feminism is bra-burning lesbianism. It's very unglamorous. I'd like to see it rebranded. We need to see a celebration of our femininity and softness," she said.

Fashion trends like lace, bows, knee socks, pigtails, baby tees, baby-doll dresses, and Hello Kitty accessories emerged to celebrate the sexy little girl. Celebrities like Baby Spice, Gwen Stefani, and Drew Barrymore expertly surfed the Girl Power crest in these fashions and bleached-blonde hair, while reports cheered their helplessness and sweet personas. By presenting not as women but as girls, their success and social capital soared, as George Washington University English professor Gayle Wald wrote in her 1998 paper "Just a Girl? Rock Music, Feminism, and the Cultural Construction of Female Youth." The "girlishly feminine persona" worn by Stefani and others "potentially furthers the notion that within patriarchal society women acquire attention, approval and authority to the degree that they are willing to act like children," Wald observed.

Instead of forcing change, Girl Power infantilized and sexually objectified its crusaders. But, as Wald foretold, the rewards for choosing the Girl Power path were great. And so, like a brigade of female Benjamin Buttons, grown women pursued the reversal of womanhood to girlhood through behavior and consumer choices.

CUTE GIRL WITH A MICROPHONE

Recording artist Lisa Loeb understands that part of her job at 90s Fest is to remind the audience who she was in the 90s. On stage, she takes requests and attempts gems so old that she can't remember the chords or key and has to reboot a couple of times. Her voice is still the girlish combination of tinkling and breathy that captivated so many in her breakup hit, "Stay (I Missed You)." It's also perfect for the children's music that has extended her career and won her a Grammy Award in 2018. She has swapped old themes of love and breakups for new ones like parents flipping pancakes.

In August 1994, twenty-seven-year-old Loeb, a Texas native, became the first artist to secure a *Billboard* number one song without a record deal. Her buddy, the actor Ethan Hawke, had finagled "Stay (I Missed You)" onto the soundtrack of the Gen X cult film *Reality Bites*. Hawke directed Loeb's video, which cemented her mainstream persona. In it, she wears a puffed-sleeve Betsey Johnson dress that Johnson herself shortened for the shoot.

"You said that I was naive / And I thought that I was strong / I thought, 'Hey I can leave, I can leave' / But now I know that I was wrong, 'cause I missed you," she croons to the camera. Her movements are charming and awkward. She shakes her wrists, loses her balance, then clutches her heart. The style of the video "created a weird intimacy" between the performer and the audience, Loeb later said. It endeared her to fans who felt they knew her, thanks in part to a heavy rotation on MTV that summer.

"Everybody loved to buy into the true story that happened to this girl, and how stars align, and sometimes it all works out," former MTV vice president of music and talent Amy Finnerty told me of Loeb's discovery. A girly image and backstory made Loeb famous, but it also contributed to a media narrative that pinned her as helpless and lucky. "I remember reading press and people would call me a waif, and I felt like I wasn't being taken seriously as a musician. That felt so strange to me, because that's what I had done my whole life: Play guitar, write music, play music. I wasn't this pop singer that appeared out of nowhere, I had been working at this forever," Loeb told *Enter-*

tainment Weekly. Hawke won credit for extracting her from obscurity. While she was grateful for the break, it stung to hear that she hadn't worked for it. "It's frustrating to read about how much I was struggling until all of a sudden Ethan got me on the soundtrack," Loeb told *Seventeen* magazine in 1995.

What distinguished Loeb from other female pop stars at the time, and helped create her girlish image, were her 60s-style cat-eye glasses accenting her otherwise bare-looking face. Her lenses, with their upswept feline corners, became a fixation. She was the toast of the near- and farsighted, and the object of countless sexy-librarian fantasies. Loeb "makes it OK to wear your nerd glasses with a sexy dress and high heels," the *Los Angeles Times* explained, permitting pretty girls to fly their inner dork flags.

Wearing glasses signaled she was educated, more so than her degree from Brown University or stint at the prestigious Berklee College of Music. Reviewers described her as "brainy," "quirky," and "intellectual." "Tortoise-shell cat-eyes frame her youthful face, giving her a piquant, slyly studious look: the rock guitarist as bohemian-intellectual songwriter," wrote the *Dallas Morning News*. The frames differentiated her from other chart-topping stars of the day who had better eyesight—or at least contact lenses. Critics found "something endearing about her out-of-place-ness" and determined that the "bespectacled Loeb offered a new range of female possibility." To many, her glasses telegraphed that she was a role model for women still attached to their girlhood and girls themselves, a smart artist for smart fans, and an antidote to the 90s girl crisis.

At the same time, Loeb's glasses also nudged her toward the schoolgirl caricature of male sexual fantasy and cliché. She "projected the demeanor of a smart college student," the *New York Times* asserted. Compared to the women who were called angry rockers, like Courtney Love, Fiona Apple, and Paula Cole, Loeb was positively virginal. *New York* magazine christened her "The Last Good Girl." "It was so much chatter about her glasses," recalled Finnerty. "I'm sure she was annoyed at some point. I mean, she never said that to me, but I'm sure she was like, 'Is anybody really listening to the song?'"

It was around this time that "the performance of girlhood," at which Loeb seemed to excel, became "a new cultural dominant within the musical practice of women in rock," according to Wald. "Acting like a girl" was advantageous for women rockers in the 90s because it offered "new ways of promoting the cultural visibility of women within rock music." The girl act owed a debt to Madonna and the Riot Grrrl movement, and allowed women to become as famous as male rockers, not by copying them but by flaunting their girliness. It also attracted a new strain of fan to market to: the teenage girl. Loeb's schoolgirl nerd chic was the perfect product.

While Loeb's girly guise launched her brand, it limited her art. As Loeb produced more music, performed widely, and gave interviews, she seemed to suggest—sometimes hinting, other times flaunting—that she was more than just a pretty, nerdy, good girl. But when Loeb pushed back against the girl trope, the media bristled. *Seventeen* wanted to know if being called a nerd offended her. "It's probably because I have glasses and I went to college and maybe because I have a lot of words in my songs. Maybe that's why they call me a nerd," Loeb said, reminding the magazine's teen audience that she was not only bespectacled but educated. Much like pioneering television shows *Sex and the*

City and *Ally McBeal*, Loeb's songs exposed women's thoughts and interior lives, giving them a platform and treating them as if they mattered. But detractors called her impressionistic lyrics "cryptic," "confusing," and "cliché." Instead, they seemed to want a sad girl singing a breakup song.

Once "Stay" reached earworm status, onlookers questioned Loeb's credibility and authenticity. Did she deserve to be the ringleader of girly nerd chic? Many asked whether or not her glasses were *real*. Did she need them to see, or were they a prop? A character on the sitcom *NewsRadio* called Loeb's glasses "fake." Soon, she was defending her frames to the press. "I'm very nearsighted," Loeb told her hometown paper, the *Dallas Morning News*. "Without my glasses, I can't see anything." She even demonstrated her spectacles' medical necessity by removing them for reporters. A talk show host invited Loeb to appear as a guest, then asked to try on her frames. Loeb's defenders emphasized that she suffered from a contact lens allergy, and that her glasses were legit because she had worn them since seventh grade.

I remember debating the realness of Lisa Loeb's glasses with my friends. A lot was at stake. If they were real, then she had a legitimate claim to unicorn status: a woman who was talented, hot, and as smart—or smarter—than men. If her glasses were fake, it meant that being sexy was the only real power, and that looks could get you further in life than being intelligent. Also, if her glasses and brainy persona were chicanery, maybe her talent was, too. Maybe she wasn't such a great songstress after all. "My glasses are a normal and real part of me. They aren't an act. I'd be selling out if I didn't wear them," Loeb said.

In addition to assailing her eyewear and intelligence, critics derided Loeb as something possibly worse—a mere cute girl with a microphone. Since she didn't play her guitar in her hit video, many thought she didn't play an instrument at all. Loeb didn't realize that she would need to "prove to everybody that I had a rock band and I had been doing this forever" to gain musical credibility, she told *Entertainment Weekly*. Now, she sees the effects of that choice on her career; it allowed people to mistake her for a lightweight. The *Boston Globe* called her a "wimpy one-shot wonder." "We remain unconvinced there's much there," the paper wrote. She was "not a great singer," after all, with a voice "alternately bratty and tender," wrote a *New York Times* music critic. Her "girlish wail and tricky lyrics suggest a precociously wise child shouting out the answers to knotty psychological questions," the assessment continued. "The glasses mean she's smart; the clothes that reveal her body mean she's hot; the hurt tone and supplicating gestures mean she's a victim," translated *New York* magazine. "We're supposed to perceive her persona as a cute little accident, but it's actually shrewdly conceived."

How did she handle being stereotyped? "I don't know. I think I'm just very persistent," Loeb told me at 90s Fest. "I was raised that way." While her subsequent albums were said to "fail to gain much traction with the pop market," Loeb has continued to make music, star in reality television shows, and perform for kids—all while wearing glasses.

Today, if you ask her about them, or about being a nerd girl before it was cool, Loeb turns spokesmodel and pounces on the opportunity to discuss her eyewear line. The alluring, girlish glasses are still her calling card, and now she sells them. "I used to prefer to talk about my music than my glasses, but now I have the Lisa Loeb Eyewear line that's available at opticians' and ophthalmologists' offices as well as at Costco," she told me. They retail for around $150. "I'm really happy to help women who wear glasses feel comfortable wearing glasses if they have to, and look pretty and maybe a little nerdy but hopefully more like a sexy librarian."

WANNABE

Comedians Lauren Brickman and Carly Ann Filbin wish the Spice Girls were performing at 90s Fest in Brooklyn. They're not, but a group of look-alikes have attracted the cameras. "I totally fell hook, line, and sinker for the Girl Power thing," Brickman says. "I loved girls cheerleading for other girls." I felt the same way in 1996, which is why I, too, loved the Spice Girls. Despite their sexualized marketing to young girls and pandering to the male gaze, I associated the Spice Girls with independence and opportunity. While most sugar pop sold a packaged, bikinied beauty ideal and the promise of male-controlled romantic love, the Spice Girls proved that girl gangs were powerful. That was what their hit "Wannabe" was all about. Here were five women who seemed to espouse female strength and sisterhood values. They made girlhood look strong and fun. Their 1996 debut, *Spice*, sold nineteen million copies in a little over a year, and is one of the bestselling records in history. Something was hitting home.

Like many girl and boy bands, the Spice Girls were Frankensteined together by a music management juggernaut. Simon Fuller, the recording executive and creator of the pop star competition show *American Idol*, launched them to stardom. The

Spice Girls became one of the most lucrative acts ever and the bestselling girl group of all time. And while "Girl Power" was the group's maxim, they were obviously cut from a pattern to attract men. Each sexy Spice Girl dressed in midriffs, animal prints, and thigh-high boots, and each fulfilled a different fetish. Sporty was the tomboy, Posh the icy model, Baby the little girl, Ginger the bombshell, and Scary the bad girl, who also happened to be the only black woman in the group. Practically every song they recorded deals with sex—the kind that men wanted and women obliged. The Spice Girls' brand extended far beyond the songs to movies, dolls, outerwear, cell phones, school supplies, and more. The Spice Girls were even superimposed on a branded mirror with which owners could attend to their mugs. The package it came in had "Girl Power" plastered all over it, implying that girls could find strength by maintaining their looks.

The softening of Riot Grrrl into Girl Power is thanks, in no small part, to the British pop band, which firmly rooted the Girl Power idea in the public consciousness. To the Spice Girls, Girl Power meant that you could wear thongs and teetering heels with your arm slung around your best girlfriend—who you cared about more than a boy, as the girls suggest in their video for "Wannabe." The Spice Girls' calling card was seductive dance moves, clothes, and lipstick. They gyrated in Lycra as if it were the ticket to kicking ass. But they also rejected the notion that they were empty fuck toys. When a radio host asked whether the women believed in sex on the first date, Melanie Brown (Scary Spice) answered, "What are we talking about sex for when we've got Girl Power to talk about?" This directness drew fans and followers who identified with wanting more than to be badgered about their sex lives. The Spice Girls fulfilled the role of hyper-sexualized female pop star while flirtatiously questioning that trope. It was a dizzying, highly successful contradiction.

Some believe that the brand of Girl Power delivered by the Spice Girls and similar vectors was good for women and girls. Former *Seventeen* magazine entertainment director Claire Connors said the Spice Girls' message was exactly what teen girls in the 90s needed. "I just thought they were the coolest band ever. I loved the Girl Power part. That was what we had been waiting for in terms of pop. It was a girl boy-band. They talked about what it meant to be a powerful girl," Connors told me. She recalled doing the first-ever Spice Girls magazine cover when she worked at *React*, which appeared as a newspaper supplement and in schools in the 90s. She's a firm believer that Girl Power via the Spice Girls did young women tremendous good. "Anything pro-girl or that makes girls happy and feel good about themselves, that's a plus. No matter what," she said.

Even the biographer of the Riot Grrrl movement, Sara Marcus, said acts like the Spice Girls kept Riot Grrrl feminism alive, even if it was a cheapened version. Brickman and Filbin agree. "I think it's really unfair to point out the Spice Girls as overmarketed or whatever," said Filbin, a spokesmodel for a shopping app. "That's antifeminist to me." "They didn't take themselves seriously," said Brickman. "It was like, yeah, they might have been sexualized, but they seemed like they were in control of it and they had humor about it."

"Back then I don't think that's why I loved them," says Filbin, scrolling through her phone. "That's why I love them now."

Bust magazine editor Debbie Stoller wasn't bothered by cor-porations adopting Girl Power as a marketing tool. "Even if this stuff gets out in the mainstream in a watered-down form, at least it's getting out there. The corporations aren't getting it quite right, but at least they're not getting it so wrong," she told the *New York Times* in 1997.

Meanwhile, Girl Power did wonders for the Spice Girls' brand and bank accounts. Between tours, albums, and prolific and diverse merch, they earned some $75 million per year at the height of their fame.

CUDDLE CORE

Girl Power wasn't only about celebrating girlhood. It also aimed to turn adult women into little girls. The wispy-haired, apple-cheeked Baby Spice personified a strand of Girl Power called Cuddle Core, which peddled girly fashions and attitudes to 90s grown-ups. Cuddle Core proclaimed there was power in acquiring the trappings of childhood. Through trending baby tees, mini backpacks, cartoon characters, sweater sets, barrettes, pastel nail polish, knee socks, and Mary Janes, infantilization became de rigueur. Cuddle Core resembled the ravished-doll look popularized by Courtney Love in her Peter Pan collars and torn tights, but cleaned up. The trend subtracted the disheveled parts and the artistic statement, and instead augmented the lace and frills. The *New York Times* called it "Love's Baby Soft feminism." During the 90s, fashion catalogue dELiA*s, jewelry shop Claire's, and girls' clothiers like the Limited Too supplied the building blocks of the Cuddle Core uniform.

This aesthetic and attitude signaled a mammoth shift. Many 90s girls had spent their formative years in androgynous flannel, skater jeans, and razored haircuts. Now, the girly girl in frilly tops, pleated skirts, and Crayola-colored eye shadow was claiming the mantle of feminism. "A lot of women can dress like little girls but be very powerful," enthused sixteen-year-old Erin, the subject of a *Los Angeles Times* article titled "Cute. Real Cute" that reported on the Cuddle Core trend. In addition to carrying a Hello Kitty character wallet and wearing a pink grosgrain ribbon in her hair, Erin admitted to owning three pairs of Mary Janes.

Retailers, the media, and the entertainment industry created and applauded the Cuddle Core trend. The Barneys New York 1997 spring fashion catalogue titled "Us Girls" featured models swathed in pink and wearing tiaras alongside copy that read, "Girls can be offended when they're whistled at but upset when they're not" and "Girls can hail a cab with their legs." When these messages were deemed sexist, Barneys defended them as celebrations of girls in a postfeminist world. *Bust* magazine made "the voice of the new girl order" their rallying cry. Butt-kicking schoolgirls like Buffy the Vampire Slayer and Sabrina the Teenage Witch figured into this trend. The Girl Power bromide appeared on paraphernalia like school supplies, jewelry, and apparel. Mountain Dew soda ran ads featuring girls doing extreme sports set to a version of the song "Thank Heaven for Little Girls" by the band Ruby. Revlon's nail polish color Girly, a juicy pink, claimed to represent "being modern, ageless and unfettered." Pink was the new power color—the new black. To embrace Girl Power was to harness the signifiers of girlhood as a better, more palatable feminism that could accord women attention and influence.

Girl Power proved successful in the marketplace, but it also popped up in government and nonprofit initiatives. The Department of Health and Human Services launched its "Girl Power!" campaign in 1996 to promote healthy behaviors for girls ages nine to fourteen, like exercising and staying drug-free. "We must teach girls that the size of their ambition is more important than the size of their clothes," wrote Health and Human Services Secretary Donna Shalala. Still, stores like dELiA*s and 5-7-9 didn't routinely stock plus sizes. Nonprofits sprang up to do Girl Power work in local communities. The Ms. Foundation's Take Our Daughters to Work Day redirected its focus on working women to boosting the self-esteem of girls. And while these efforts seemed worthy, they were inadvertently tying girls' self-worth to the pursuit of perfecting appearance and amassing possessions.

JUST A GIRL

Gwen Stefani's song "Just a Girl" tapped into the ethos of Girl Power as a bridge to self-esteem. When rock critic Kim France profiled Stefani for the January 1997 issue of *Us* magazine, she went to see Stefani's band, No Doubt, perform in Los Angeles. "I remember sitting in the stadium and she's chanting to the girls in the audience. And there are schoolgirls there. Young girls. And she's going, 'Fuck you, I'm a girl!' And they're going, 'Fuck you, I'm a girl!' I thought, this is going to be a formidable generation growing up with that message. It was so different than the message I grew up with," she told me.

The hit that made Stefani, "Just a Girl," challenged and mocked the adages that girls should be polite and demure. "'Cause I'm just a girl, a little ol' me / Well don't let me out of your sight," she croons. "Oh I'm just a girl, all pretty and petite

/ So don't let me have any rights / Oh . . . I've had it up to here!"
The teen girls France described who idolized Stefani helped
launch her to a superstardom that continues more than two decades later. Girls latched on to how Stefani indicted the helpless,
delicate girl stereotype while embodying it in the same breath.
Spin magazine described her fans—the Gwennabees—as "smatterings of breathlessly excited, blonde-streaked, sparkle-lashed
14-year-olds" who "litter the backstage area."

Performing "Just a Girl" in the 90s, Stefani was sarcastic
and implacable. She seemed to cover the entire stage at once,
scaling speakers, cartwheeling, kicking and throwing her arms
like snakes sprung from cans, adorned in track pants, a crop top,
and a bindi. She was not afraid to sweat, and even fractured a
foot during one concert. One reporter claimed that she offered
him her bra and shirt to smell after a particularly aerobic performance to prove that she was *such* a girl that excessive sweating
didn't make her smell bad.

Stefani's ode to girldom marked an end to the "petulant
whining" of plain-faced rocker women who "openly spew their
feelings," according to a 1996 *Entertainment Weekly* review.
There was palpable excitement in the ouster of unshaved Paula
Cole, damaged Fiona Apple, and trashy Courtney Love. In place
of such artists was Stefani, a "camera-ready" blonde goddess who
looked like "a cross between Jessica Lange and a naughty cheer-
leader," observed the *Los Angeles Times*. She was compared to
"larger-than-life celluloid sex symbols" like Pamela Anderson,
Jenny McCarthy, and Anna Nicole Smith. Stefani's "pouty" voice
was a "girlish hiccup"; her lyrics were "mewling" and "worthy of
the rescue-me blankness of Mariah Carey's entire repertoire,"
EW continued. The glad-to-be-ogled girly girl was back. Re-

viewers cast Stefani as a sex kitten, "the paragon of baremidriffed yumminess." She had the blonde, thin body men idealized while still looking like she belonged in high school. Fittingly, reports in news and music publications focused on Stefani's passion for makeup, boys, and other elemental accoutrements of girlhood. "If it's pretty, she wears it; if it smells good, she sprays it, and if it's feminine, she flaunts it," explained the *Chicago Sun-Times*. "I love makeup. I love getting my hair done. I love pedicures. I'm the furthest thing from a rock chick," Stefani told *Spin*.

Critics infantilized Stefani, making her seem less like a rising rock star than an average teen girl in Peoria, even though she was in her midtwenties. Their proof was all the ways she deferred to her mom and dad. Before Stefani got famous, her parents forbade her from playing late gigs far from home on school nights. Her mother stopped speaking to her for a week after Stefani led a stadium in a "Fuck you, I'm a girl" chant during a show. Stefani still lived at home when No Doubt was launched out of neo-ska obscurity and into mainstream pop fame with their 1995 album *Tragic Kingdom*. She approached her father with a hand outstretched for money before rehearsals. "I don't pay bills. I don't pay rent. The only thing I pay is my phone bill and my car insurance," Stefani told a reporter. "Just a Girl" was inspired by Stefani's dad yelling at her for staying out too late at her boyfriend's house. Even after moving out of her parents' home, Stefani often returned there to sleep in her childhood bed. This veil of dependence on her parents secured her helpless "girl-next-door" status.

When she wasn't applying makeup or deferring to her dad, Stefani was mooning over her ex-boyfriend, No Doubt's bassist, Tony Kanal. Critics harped on their breakup, which was a subject of *Tragic Kingdom*. Stefani was couched as a combustible, heartbroken little girl to Kanal's dispassionate heartbreaker. The press quoted him on the band's Jamaican pop influences and quoted her about their split. "I forced Tony to go out with me," Stefani said. "He wasn't even interested. When we made out that first night I think he thought it was more of a one-night kiss. But then we started going out and after the first year, I was going, 'When are we getting married?'"

To be sure, Stefani and Kanal amped up the drama at shows. "Don't Speak" was another runaway hit in part because it detailed their breakup. She would sulk during an instrumental interlude in "Just a Girl," and Kanal would hand her a bouquet of flowers that she'd accept, then tear apart, sending the crowd into a fit. Even if it was a performance, critics and fans cheered girlhood's return. They had missed it.

Tragic Kingdom's success—more than ten million copies sold by the decade's end—reaffirmed that "sex still sells, even when it comes to women musicians," wrote *Entertainment Weekly*. "Maybe we don't live in such progressive times after all." Reviewers both celebrated the reemergence of the girly girl and skewered her for being retrograde. This is the trajectory of a 1996 *Spin* cover story about No Doubt and their hit album, which mostly relished what a girl the singer who critiqued girlhood turned out to be. She may have "embraced the girly shit," as France says, from a broken heart to makeup brands, but her punishment, like Lisa Loeb's, was that the girly girl was all she could be.

BUBBLEGUM UNICORNS

Girl Power was leveraged to sell more than just record albums. Major marketers like Coke, Toyota, Frito-Lay, and Sears "recognized the muscle of girl power," and flocked to hook teen girls early. Market research in the 90s showed that younger demographics held tremendous purchasing power. Children ages four to twelve commanded nearly $15 billion in annual sales and influenced some $160 billion in household purchases, according to a 1994 study by Texas A&M University. Leading retailers began to market to a group too young to even secure a credit card. Some 40 percent of national marketers boasted kid-specific marketing strategies. Companies started conducting research to learn just what youngsters—and girls in particular—were spending their money on. It turned out, in 1997, teen girls spent $50 billion on things like clothes, jewelry, beauty products, entertainment, and food alone. As a CNN reporter put it: "You are looking at the future of retail, and it can barely see over the counter."

Lisa Frank was known for school supplies and paper goods featuring kittens, bubbles, and unicorns in the colors of an exploded gumball machine. Lisa Frank understood Girl Power's marketplace potential and devoted ample resources to capturing it. "It's almost as if they had radar and were able to go into the minds of these young girls," said one marketing expert on CNN in 1998. What began as paper goods carried in mall favorite Spencer Gifts grew into more costly items—sneakers, outerwear, CD-ROMs, and much more. "If a little girl uses it, chances are Lisa Frank makes it," enthused the segment's host.

The company might have been prescient, but it also did its homework. Lisa Frank planted product testers in elementary schools. They recruited classrooms to focus-group experimental characters and fabrics, and created an online fan club and website. By 1998, Lisa Frank was banking $250 million a year by "catering exclusively to the whims of four- to twelve-year-old girls," according to one report. Company president James Green, who was also married to the founder and namesake, attributed the brand's success to his wife being "the world's greatest shopper" and "a little girl at heart." But their methods weren't always aboveboard. In 2001, the company agreed to pay a $30,000 fine for collecting personal information from children—such as their names, addresses, phone numbers, and favorite colors and seasons—without parental consent.

In 1994, Lisa Frank commissioned a study of eight- to twelve-year-old girls' purchasing preferences. Nearly half said the clothing section was their first stop in their favorite stores. Girls spent 41 percent of their money on clothes and shoes, 20 percent on magazines and books, 14 percent on makeup, 12 percent on jewelry, and 11 percent on school supplies. Green said the study was intended to help the company develop "product lines kids want, not just need." By 1997, teen spending reached $122 billion. Girls shelled out the most for objects promising to enhance their appearance and their self-image.

Girl Power had morphed from rebellion into a miasma of consumer ideals. It became so embedded in American consumer culture that even antifeminist offenders began to claim it to move their products. One example is a Candie's perfume ad featuring Sugar Ray's Mark McGrath. He is staring lecherously at the camera while *She's All That* star Jodi Lyn O'Keefe sits atop his computer monitor, which shows a rocket going off between her legs. Advertising Women of New York selected it as the most degrading, insulting, and reductive advertisement of the year. But Candie's chairman Neil Cole claimed that the ad was all about Girl Power. A luxury watch brand used the idea to introduce young women to expensive timepieces. *USA Today* reported that a spate of new electronic toys created for girls empowered them. But instead of challenging girls, the toys and games allowed them to check horoscopes, "swap gossip," and "pass notes" to one another.

That Girl Power was co-opted to sell products didn't surprise parents. The vast majority—92 percent—said they were worried that kid-centered television advertising was making their children "too materialistic" and "turning them into chronic consumers," according to one report. Some skeptics did push back on commodified Girl Power. "An ideology based on consumerism can never be a revolutionary social movement," said North Carolina State University professor Amy McClure in a paper about Girl Power that she presented at the American Sociological Association. "The fact that it appears to be a revolutionary movement is a dangerous lie that not only marketers sell to us but that we often happily sell to ourselves."

A VIRGIN TO BE DEFLOWERED

Kim Gordon, bassist and guitarist for Sonic Youth, thought Girl Power's most prominent and powerful emissaries, the Spice Girls, were doing more harm than good, and said so in the press. "I think they're totally ridiculous. Something out of Disneyland . . . They're masquerading as little girls. It's repulsive," she said. Out swept the tide of alternative women rockers who wrote their own songs and told their stories, who were called bitches for being angry and broken and strong. In washed a cadre of "Svengali-produced" blondes, as Paula Cole put it, manipulated and packaged for the new teen market. The sounds of Fiona, Paula, and Courtney—and the empowerment of feel-good, woman-charged

Lilith Fair—were trampled by society's ideal of what women pop stars should be: young, blonde, and, if not singing about sex, selling it. The Spice Girls pioneered this convention; Britney Spears perfected it.

The Spice Girls and Gwen Stefani seemed to casually pretend that Girl Power wasn't sexual, but Britney Spears did nothing of the sort. Her provocative 1999 *Rolling Stone* magazine cover at seventeen years old promised to take readers into her bedroom. Spears reclines in a black bra on a satin coverlet. She holds a phone with a curled cord, and hugs a purple stuffed Teletubby doll. "Inside the Heart, Mind & Bedroom of a Teen Dream," offers the cover line. This was worlds away from Alanis Morissette's 1995 cover, emblazoned with the words "Angry White Female." In her imaginary childhood bedroom, replete with florals and china dolls, Spears conjures the hypersexual, bionic future of girlhood. Brandishing her teenage sexuality was indeed powerful—abstinence be damned—except that she wasn't in the driver's seat. The monolithic music industry replaced authentic women rockers with willing teen pop princesses and told us we could expect the same powerful feminist idols.

The video for Spears's first hit, with its boundary-pushing ellipsis, ". . . Baby One More Time," panders to its prepubescent audience and is set in territory familiar to teens—the locker-lined hallways and gymnasiums of high school. Spears dons an eroticized Catholic schoolgirl uniform. A white shirt is knotted above her navel and a plaid skirt reveals her upper thighs. There are pom-poms in her pigtails. She squats low and grinds her hips, less like the Mouseketeer she once was and more like a participant in the step (or strip) club.

This presentation was thrilling for men and girls alike.

Spears talked to teen girls on their terms. She projected empowerment and freedom with every neck whip, thigh dip, and seductive gaze at the camera. Because she exuded such confidence, critics accused her of manipulating fans with sex. Take the opening of the *Rolling Stone* profile. Spears "extends a honeyed thigh across the sofa." She wears a logo T-shirt "distended by her ample chest" and silky white shorts that "cling snugly to her hips" as she "cocks her head and smiles receptively." But then, the writer shames the reader for falling into Spears's "trap," which is "carefully baited by a debut video that shows the seventeen-year-old singer cavorting around like the naughtiest of school girls." Spears is to blame for tempting men and presenting a sexually aggressive role model to girls.

Not long after the peak of her success "the pop sensation with an eleventh-grade education" suffered a psychotic breakdown. When she did her *Rolling Stone* "comeback" cover in 2008, the magazine accused her of engineering her persona. Spears was smart enough to know what the culture wanted from her, and was "created as a virgin to be deflowered before us, for our amusement and titillation," the magazine observed.

Curiously, the media blamed the onslaught of Britney, Christina Aguilera, and the Spice Girls on kids and teens, who were said to be driving the purchasing culture. But they weren't industry executives, A&R guys, or MTV producers. They were cultural consumers buying what was sold. . . . *Baby One More Time* became a top-selling album, and Spears took two categories in *The Guinness Book of World Records*—Best Selling Album by a Teenage Solo Artist and Best Selling Album in the US by a Female Artist. Any remaining Riot Grrrls vomited into their combat boots. Even watered-down Girl Power had all but evaporated, and in its place was the blatant selling of sex. Britney wasn't performing girlhood and challenging it for accolades, like Stefani. She was submitting to it, whole cloth, because it gave her wealth and power.

RAUNCH CULTURE

Girl Power transitioned seamlessly—or, rather, devolved horribly—into the raunch culture of the late 90s and early 2000s, in all of its stripper-pole workout, belly-button piercing, Playboy Bunny–iconographed glory. Women performers had perfected branding themselves as Playmates and lad mag cover models so well that "women now want to be Maxim babes as much as men want Maxim babes," wrote Maureen Dowd. "So women have moved from fighting objectification to seeking it."

Nothing symbolized raunch culture more than the *Girls Gone Wild* franchise, which launched in 1997. Producers famously trained cameras on college parties, spring break, and Mardi Gras, begging regular girls to go wild by flashing their breasts, making out with other women, or masturbating for the camera. These acts would later appear in *Girls Gone Wild* films for sale, and late-night infomercials selling them. Becoming overtly sexual could earn women cultural capital, the series promised. "Our mothers were 'burning their bras' and picketing *Playboy,* and suddenly we were getting implants and wearing the bunny logo as supposed symbols of our liberation," wrote Ariel Levy in *Female Chauvinist Pigs,* her meticulous 2005 account of raunch culture. As Levy documents, women participated actively in this movement. They were editors at *Maxim,* took striptease classes, and bought vampy stilettos. In 2007, the Spice Girls performed at the Victoria's Secret Fashion Show, reinforcing that what Girl Power was actually about was girls' freedom to exploit themselves.

As raunch reigned and Spears swayed in her shrunken clothes, an epidemic of total pubic hair removal was sweeping the country's young women. The Brazilian bikini wax appeared as yet another solution to marketers' profligate assertions that women were too fat, smelly, and hairy. It was also the seeming preference of pornographers. Unsurprisingly, young women had internalized that pubic hair was unclean and unsightly. Even though it was something most people never saw, instead of just hiding it in shame, young women were offered a fix: get rid of it altogether. What differentiated the Brazilian from prior bikini wax types was the removal of not only all pubic hair, but also all buttocks hair. Eve Ensler's feminist celebration of women's anatomy, the 1996 play *The Vagina Monologues*, was particularly critical of men who requested that women remove their pubic hair. But women began to report that they weren't doing it for men, but for themselves. It was yet another market manifestation of Girl Power—the backward slide of women enduring great pain and paying good money to look like little girls.

The seven Brazilian Padilha sisters opened the iconic J Sisters salon in Midtown Manhattan in 1987, which is by many accounts the first practical and ideological home of the Brazilian bikini wax in America. The *New York Times* first mentioned it in 1998, when a reporter sought out the most painful of beauty treatments. *Salon* covered the procedure's popularity among celebrities in 1999. Kirstie Alley said of her wax, "It feels like a baby's butt, only all over," and Gwyneth Paltrow thanked the

J Sisters by simply saying, "You've changed my life." In 2000, Carrie Bradshaw felt in charge when she waxed it all off in *Sex and the City*, which made the "Manolo Blahnik demographic sit up and take notice." Brazilian waxing soon became compulsory for coeds and young women.

Fear of pubic hair and the desire to remove it persists. By 2011, Indiana University sex researchers Debby Herbenick and Vanessa Schick found 60 percent of women eighteen to twenty-four went bald eagle. Women under thirty were two to three times as likely to do it as women over thirty. A 2016 study found that of the women who removed all of their pubic hair, 59 percent claimed they did so for hygienic reasons. The idea that pubic hair is unclean is a complete fallacy, according to doctors consulted about the study.

Indeed, over the course of the 90s, the Riot Grrrl insurrection was similarly ripped out at the roots—slowly at first, then steadily, then rapidly—by Girl Power marketing. By the end of the millennium, however, being girly was out, and in its place was the promise to girls and women that they could be empowered only by turning themselves into sex objects. The media, society, and Hollywood no longer needed to do it for them. They could do it to themselves. And also to their children.

In 2002, controversy-loving mall staple Abercrombie & Fitch introduced thongs into its children's clothing line. Christian and family groups protested the underwear, which read "wink wink" and "eye candy," and were made to fit seven- to fourteen-year-olds. Abercrombie & Fitch would not be dissuaded from marketing sex to prepubescent girls and their families, however, and refused to pull the thongs from shelves. President Mike Jeffries has since defended his thongs for children this way: "People said we were cynical, that we were sexualizing little girls. But you know what? I still think those are cute underwear for little girls. And I think anybody who gets on a bandwagon about thongs for little girls is crazy."

EPILOGUE

I started researching *90s Bitch* within days of Hillary Rodham Clinton announcing her second bid for the US presidency, in April 2015. In contrast to her 2008 campaign, Clinton was the undisputed front-runner—not only for her party's nomination, but for the presidency itself. Throughout the course of the 2016 election cycle, it was widely believed that the first woman to ascend to the White House—to break that "highest, hardest glass ceiling"—would propel all women forward. Once again, a new era of gender parity was promised by the fact of Clinton's historic candidacy and assumed victory, and also in the themes and talking points of her campaign. Whereas Clinton addressed gender inequality somewhat cautiously in 2008—and played down the significance of a woman running for the presidency—she confronted it head-on in 2016. Clinton was often heard on the campaign trail saying, "If playing the gender card means fighting for equal pay and women's healthcare . . . Deal! Me! In!"

Many thought 2016 would be different—that society had changed dramatically since the 1990s, when Clinton was first pilloried for her comments about baking cookies and Tammy Wynette, and even since her 2008 presidential bid. Indeed, much had changed. By 2016, women accounted for nearly half of the

American workforce, and their median hourly wage had climbed 25 percent in the previous thirty years. Millennial women bested their male counterparts in higher education—more women than men were likely to enter college, and more women than men ages twenty-five to thirty-two held bachelor's degrees. These and other markers led a 2013 Pew Research Center study to declare, "Today's young women are the first in modern history to start their work lives at near parity with men." Even on Capitol Hill, more women are running for and winning political office today than ever before. In 2017, a total of 104 women served in the United States Congress—twenty in the Senate and eighty-four in the House of Representatives—up from just four and twenty-nine respectively during the Year of the Woman in 1992.

Girls' lives have undoubtedly improved, as well. After peaking in the 90s, rates of teen pregnancy, birth, and abortion have plummeted to historic lows thanks to expanded access to birth control, among other factors. In 2016, a monumental study from Massachusetts General Hospital that followed women diagnosed with eating disorders for twenty years or more found two-thirds of them get better. These difficult-to-treat diseases might not be the "life sentence" clinicians once thought they were.

The rise of the feminist internet has elevated new voices and issues, and turned a magnifying glass on mainstream inequities and institutional oppression in ways unthinkable during the 90s. Know Your IX educates students about how to report sex crimes. Hollaback! combats street harassment. Awareness campaigns like #GirlsLikeUs promote transgender recognition. #YesAllWomen highlights gender violence. #MeToo unites victims of sexual harassment and assault, and #TimesUp established a legal defense fund for victims. Black Lives Matter, the movement to end systemic racism, was founded by women. All of these efforts have grown and thrived online. Leading entertainers and politicians have identified with feminism loudly and proudly, signaling that fighting for gender equality is not only vital but vogue. And, of course, a majority of Americans voted for a woman president.

I finished writing this book just as Clinton conceded the presidential campaign to Donald Trump. Clinton's loss was caused by a multitude of factors, but it delivered an undeniable blow to American women. Many likened the feeling to grieving for a loved one. Indeed, the mood at Clinton's election night party at the Javits Center's River Pavilion in Manhattan resembled a funeral. The campaign chose the venue because of its 81-foot high glass ceiling, which Clinton planned to metaphorically shatter that night. The evening of returns-watching began with gaiety, excitement, and jolly drinking. It ended with increasingly heavy drinking dampened by tears and sobs. A couple of lucky little girls were peppered throughout the room packed shoulder to shoulder with adults. One of them, probably seven or eight years old, clung to her mom's pant leg. Another wrapped herself in her father's arms. They both studied the grown-ups' faces, the television screens, and the dirty carpet for some indication of what was going on. This was supposed to be a happy, history-making night. Men and women huddled in corners and on the floor, making the space feel even more cavernous, the ceiling even further from reach.

Trump's victory and his presidency not only shocked a nation convinced that Clinton would win—it shocked those who were certain that gender parity might finally be at hand. Maybe that was always wishful thinking. But the misogynistic tenor of the election and what passes muster in the current administration shouldn't have surprised us. History merely repeated itself, as it so often does. Just as the sparkling promise of women's accomplishments in the 80s faded in the bitchified 90s, Hillary Clinton, the front-runner and herald of gender parity, was stopped by a rival campaign of crude sexism. Today more than ever, many wonder whether anything fundamental has changed for American women since the 90s or, like the mirage of a woman president, it just appeared as though it had.

Since the 90s, some positive trends have actually come to a halt, or reversed themselves. The US maternal mortality rate—the number of women dying during pregnancy, childbirth, and the early weeks postpartum—is rising. Women have made progress at work but still fall short of parity on many levels. One study of women working in film showed no growth at all. Women were only 17 percent of writers, directors, producers, and cinematographers of the 250 top-grossing movies in both 2016 and 1998. More women work in television today than in the 90s, but their ranks are still small. The percentage of women who work behind the scenes in broadcast programs has grown only six percentage points in eighteen years. Teen births have decreased markedly, but the United States still has among the highest rates in the developed world.

Women today are dogged at every turn by sexism. The gender wage gap still exists. While women's median hourly wage has risen, it's only 84 percent of men's. America is the only industrialized nation lacking mandatory paid parental leave.

Moreover, women are hired and promoted at lower rates than men, according to a 2016 study by LeanIn.org and McKinsey & Company that looked at data from 132 companies employing more than 4.6 million people. For every 100 women promoted, 130 men rise in the ranks. Sexism's economic impact on women is only beginning to be discussed and understood.

Access to quality healthcare and family planning services are increasingly stymied, particularly hindering low-income, minority, and rural women. College campuses are where women can feel the most equal to men, but they can also be dangerous places—43 percent of dating college women report violence or abusive behavior from those they date, while 19 percent of college women report being sexually assaulted. These figures are inherently conservative, since rape and sexual assault are notoriously underreported no matter where they occur. And they occur everywhere.

In our increasingly pornified culture, misogynistic stunts and talk are the norm. We see and hear them nightly on reality television and cable news, or online. They pervade mainstream advertising and Hollywood, where women's bodies are blatantly for sale. "Locker room talk" and attitudes underpin the rape culture typified by the Toronto police officer who, in 2011, counseled women that to avoid rape they should not dress like "sluts," and the father who, when his son was accused of rape in 2016, defended him by insisting he should not to go to prison for "twenty minutes of action." Male entitlement to sex with women fueled Santa Barbara shooter Elliot Rodger's slaughter of six people. Powerful men in media and Hollywood like Harvey Weinstein, Charlie Rose, Matt Lauer, and countless others who shape societal narratives and decide what's news have preyed on women with impunity for years. Eighty-one percent of women have experienced sexual harassment according to a report released in early 2018. Sixty-six percent said it occurred in public spaces and 38 percent said it happened at work. Maybe we shouldn't be so surprised that the blatant and lewd sexism woven into the fabric of our society finally emerged unabashed in a modern presidential campaign, and in the White House itself.

Putting aside the obvious differences, the 1990s and the 2010s actually have much in common. In the 1990s, the onset of the twenty-four-hour news cycle changed American politics and culture forever. It also negated hard-fought gains made by pioneering women and undermined feminism at large. As this book documents, the glaring commonality shared by 90s women who sought influence is that, instead, they became infamous. They were perpetually, incessantly bitchified for threatening established male power, or for simply showing up in public.

In the 2010s, American politics and culture have been upended once again by a communications revolution—social media—and women remain in the crosshairs. This time, however, they are targeted not only by content publishers, but also by anyone in America with an itchy trigger finger. Forty-one percent of American women have been sexually harassed online. Women experience a wider and more dangerous variety of harassing behaviors online than men, and are more likely to self-censor what they post in hopes of avoiding abuse. We hardly need studies to alert us to this fact. Indeed, the established norm is not so much online harassment as rampant chauvinism, which is as evident online as in real life.

The 90s was a decade obsessed with sex, and women who

came of age in that decade internalized a culture of sexism, both blatant and subtle, social and commercial. We didn't question the frame because we didn't know that we should. But now we do—and that's one thing that has changed since the 90s. We now know not to be surprised by today's misogyny, because it was seeded and cultivated decades ago when bitchifying any woman, every woman, was just the way things were. Knowing this history is how we stop it from repeating. We can actually put our 90s nostalgia to potent use.

Let's no longer assume, as we did in the 90s, that more women attending college, marrying later, running companies, and holding major government positions is enough to ensure our humanity alongside men. Instead, let's reexamine the stories that are told and sold about women—that we tell and sell ourselves. Probing the failed promise of gender equality for truth and meaning is the first, essential step in confronting the sexism that suffuses women's lives today—and to prevent it from suffusing the lives of our daughters and sons in decades to come.

ENDNOTES

PROLOGUE

ix **by calling them dogs in heat:** Clare Bayley, *Bitch: A History*, June 2, 2011, http://clarebayley.com/2011/06/bitch-a-history/.

INTRODUCTION

xiii **they earned more bachelor's, master's, and associate's degrees than men:** *Women in America: Indicators of Social and Economic Well-Being*, US Census Bureau, March 2011, https://www.census .gov/library/publications/2011/demo/womeninamerica.html.

xiv **the median marriage age for women swung between twenty and twenty-two, but in 1990, it nearly jumped to twenty-four:** *Women in America*, US Census Bureau.

xiv **reached twenty-five:** *Women in America*, US Census Bureau.

xiv **women accounted for close to 30 percent of lawyers:** Peggy Orenstein, *Flux: Women on Sex, Work, Love, Kids, and Life in a Half-Changed World* (New York: Random House, 2000), 4.

xiv **"Women are the new providers":** Tamar Lewin, "Women Are Becoming Equal Providers," *New York Times*, May 11, 1995, http://www.nytimes.com/1995/05/11/us/women-are-becoming -equal-providers.html.

xv **"They were a bunch of nerds":** Coolio, interview with the author, September 2015.

xvi **after meeting at the liberal arts school Adelphi University on Long Island:** S. Craig Watkins, *Hip Hop Matters: Politics, Pop Culture, and the Struggle for the Soul of a Movement* (Boston: Beacon Press, 2006), 114.

xvii **"throw tantrums in Bloomingdale's":** Elizabeth Wurtzel, *Bitch: In Praise of Difficult Women* (New York: Anchor Books, 1999), 30.

xvii **"I'm tough, I'm ambitious, and I know exactly what I want":** Mary Biggs, *Women's Words: The Columbia Book of Quotations by Women* (New York: Columbia University Press, 1996), 187.

CHAPTER 1: PRETTY ON THE OUTSIDE

1 **Americans didn't just watch—they binged:** Douglas Kellner, "The Persian Gulf TV War Revisited," UCLA, https://pages .gseis.ucla.edu/faculty/kellner/papers/gulfwarrevisited.htm.

1 **1991 Times-Mirror survey:** Kellner, "The Persian Gulf TV War Revisited."

2 **"beautiful, dependent, helpless . . .":** Mike Feinsilber, "TV Poorly Serves People Who Rely on It Most, Report Says," Associated Press, February 25, 1992.

2 **report by an American Psychological Association task force:** Aletha C. Huston et al., *Big World, Small Screen: The Role of Television in American Society* (Lincoln: University of Nebraska Press, 1992).

3 **The proportion of women characters appearing on prime-time television barely budged:** Tom Tapp, "New Survey Says Tube Not Tops for Women," *Daily Variety*, September 17, 1999.

3 **15 percent of the creators of the top hundred primetime shows:** Tapp, "New Survey Says Tube Not Tops for Women."

3 **The program's credits:** Susan J. Douglas, *The Rise of Enlightened Sexism: How Pop Culture Took Us from Girl Power to Girls Gone Wild* (New York: St. Martin's Press, 2010), 28.

3 **close to eleven million households:** Steven Herbert, "Fox 'Hills' Strategy Pays Off," *Los Angeles Times*, July 13, 1991,

http://articles.latimes.com/1991-07-13/entertainment
/ca-1884_1_beverly-hills.

3   **half of teenage girls polled:** Bruce Horovitz, "Marketers Re-
    think Show's Teen Appeal," *Los Angeles Times*, December 22,
    1992, http://articles.latimes.com/1992-12-22/business/fi-2407
    _1_teen-talk.

6   **A thirteen-year-old Manhattan private schooler revealed:**
    Anne Jarrell, "The Face of Teenage Sex Grows Younger," *New
    York Times*, April 2, 2000, http://www.nytimes.com/2000/04/02
    /style/the-face-of-teenage-sex-grows-younger.html?pagewanted
    =all.

7   **Nearly 70 percent of elementary school girls:** A. E. Field et
    al., "Exposure to the Mass Media and Weight Concerns among
    Girls," *Pediatrics*, March 1999, https://www.ncbi.nlm.nih.gov
    /pubmed/10049992.

8   **girls were nearly twice as likely as boys to believe they were
    overweight:** National Center for Health Statistics, "Many Teens
    Engage in Risk-Taking Behaviors That Can Lead to Chronic
    Disease, Injury, or Death," news release, July 11, 1995, https://
    www.cdc.gov/nchs/pressroom/95facts/fsyrisk.htm.

9   **more than 40 percent of first through third graders wanted
    to be thinner:** Elizabeth Collins, "Body Figure Perceptions and
    Preferences among Preadolescent Children," *International Jour-
    nal of Eating Disorders* 10, no. 2 (March 1991): 199–208.

9   **Cultural aspiration to thinness could explain:** Mandy Mc-
    Carthy, "The Thin Ideal, Depression and Eating Disorders in
    Women," *Behavioral Research and Therapy* 2, no. 3 (1990): 205–
    215, http://www.darkcoding.net/research/the_thin_ideal.pdf.

9   **"a strong blast from a blow dryer would waft her away":**
    Louise Lague, "How Thin Is Too Thin?" *People*, September 20,
    1993, http://people.com/archive/cover-story-how-thin-is-too
    -thin-vol-40-no-12/.

10 **Women looked at thinning female bodies:** Tracie Egan
    Morrissey, "Portia de Rossi, *Ally McBeal*, and a Generation of
    Eating Disorders," *Jezebel*, November 2, 2010, http://jezebel
    .com/5679536/portia-de-rossi-ally-mcbeal-and-a-generation
    -of-eating-disorders.

10   **"assumed a new presence in the lives of Americans"**: Keith Bradsher, "A Job for Real Men: Buying Lingerie," *New York Times*, February 4, 1990.

11   **"When I tried to buy lingerie for my wife"**: Merrill Fabry, "The History Behind 'Victoria' in Victoria's Secret," *Time*, December 8, 2015.

11   **nearly six hundred stores**: Stephanie Strom, "Profile: Grace Nichols; When Victoria's Secret Faltered, She Was Quick to Fix It," *New York Times*, November 21, 1993.

11   **"what was once a discreet (or salacious) business"**: Bradsher, "A Job for Real Men: Buying Lingerie."

12   **By 1992, more than $51.5 billion:** Holly Brubach, "Mail Order America," *New York Times*, November 21, 1993.

12   **By 1997, Victoria's Secret was shipping out 450 million catalogues:** Casey Lewis, "The Rise and Fall of the Victoria's Secret Catalog," *Racked*, July 25, 2016, http://www.racked .com/2016/7/25/12119174/victorias-secret-catalog-rip.

13   **Victoria's Secret reached $1 billion in revenue:** Strom, "Profile: Grace Nichols."

14   **one report likened its gain on the panty:** Asra Q. Nomani, "How Thong Underwear Managed to Win Over a Mainstream Market," *Wall Street Journal*, June 8, 1999, http://www.wsj.com /articles/SB928797031306049284.

14   **"the fastest growing segment of the $2 billion a year women's panty business"**: Nomani, "How Thong Underwear Managed to Win Over a Mainstream Market."

14   **"You lifted the back of your jacket"**: Monica Lewinsky, interview by Barbara Walters, ABC, March 1999.

15   **"the garment that shook a presidency"**: Nomani, "How Thong Underwear Managed to Win Over a Mainstream Market."

16   **more than two billion people:** Brendan Baber and Eric Spitznagel, *Planet Baywatch: The Unofficial Guide to the New World Order* (London: Michael O'Mara Books, 1996).

17   **"sizzling sweet eye candy"**: Leah Ollman, "Galleries: Art That's Instantly Gratifying," *Los Angeles Times*, May 28, 2004, http://articles.latimes.com/2004/may/28/entertainment/et -galleries28/2.

17 **"out-there sex weirdo":** Linda Stasi, "After 11 Years Looking at This: Behind Baywatch Reveals the Interesting Side," *New York Post*, June 29, 2001, http://nypost.com/2001/06/29/after -11-years-looking-at-this-behind-baywatch-on-e-reveals-the -interesting-side/.

17 **in small outfits:** John Crook, "Pamela Anderson Makes Her Bow as Sitcom Star in 'Stacked,'" *St. Louis Post-Dispatch*, April 10, 2005.

17 **"erotic cartoons":** Michael Wilmington, "This 'Naked Gun' Misfires as It Runs Out of Ammunition," *Chicago Tribune*, March 18, 1994, http://articles.chicagotribune.com/1994-03 -18/entertainment/9403180278_1_frank-drebin-david-zucker -academy-awards-show.

17 **"freakish mammary glands":** Patt Morrison, "Haughty, Naughty: A 'Model' Word," *Los Angeles Times*, November 29, 1995, http://articles.latimes.com/1995-11-29/local/me-8377_1 _modeling-agency.

17 **were a primary subject:** "RARE Anna Nicole Smith Interview Regis Philbin 1992 GUESS JEANS," YouTube video, 6:26, posted by Guess Again, May 27, 2014, https://www.youtube .com/watch?v=6C-u5J6oqP4.

17 **"Did you have breast augmentation?":** Anna Nicole Smith, interview by Larry King, CNN, May 29, 2002, www.cnn.com /TRANSCRIPTS/0205/29/lkl.00.html.

17 **"Everything I have is because of them":** "Obituary: Anna Nicole Smith," *Economist*, February 15, 2007, http://www .economist.com/node/8697358.

18 **"the number one requested body image in L.A.":** Dara Welles, "Teamsters Say They're Ready to Get Back to Work: Dow Chemical May Face Lawsuits in Reference to Breast Implants," CNN, August 19, 1997.

18 **Godiva chocolates at every photo shoot:** Matthew Heller, "The White Widow," *The Independent*, February 11, 1996, http://www.independent.co.uk/arts-entertainment/the-white -widow-1318392.html.

19 **"uncontainable by ordinary clothes":** Philip Kennicott, "The Fantasy of Happily Ever After," *Washington Post*, February 9, 2007.

19   **"She brings back visions of Hollywood glamour":** Susan
Schindehette and David Hutchings, "Anna Nicole Smith Models
for Guess Jeans," *People*, April 12, 1993, http://people.com
/tbd/from-the-archives-anna-nicole-smith-models-for-guess
-jeans-1993/.

19   **"She spilled out of her tops":** Kennicott, "The Fantasy of Happily Ever After."

19   **"There were only two of them":** "Obituary: Anna Nicole
Smith," *Economist.*

20   **"Something is going on":** Ellen Crean, "What's Up with Anna
Nicole Smith?" CBS, December 30, 2004, http://www.cbsnews
.com/news/whats-up-with-anna-nicole-smith/.

21   **"thought they would shake things up":** James Kahn, interview with the author, October 2015.

21   **"bothersome Issues and Morals":** David Wild, "'Melrose
Place' Is a Really Good Show," *Rolling Stone*, May 19, 1994.

21   **Actress Hunter Tylo sued:** "Would-Be 'Melrose' Actress Wins
Nearly $5 Million Award," CNN, December 22, 1997, http://
www.cnn.com/SHOWBIZ/9712/22/melrose.lawsuit/.

22   **"stop-at-nothing female character":** Denise Gellene, "Ad
Agency Women Hit TV Stereotypes: From 'Bewitched' to 'Melrose Place,' Life on Madison Avenue Has Been Distorted, Critics
Say," *Los Angeles Times*, May 20, 1996.

22   **"she is really very nice, the kind of woman":** Elizabeth Kolbert, "So Nice to Be Mean," *New York Times*, April 3, 1994.

22   **"I love it":** Bruce Fretts, "Entertainer 4: Heather Locklear," *Entertainment Weekly*, December 30, 1994, http://ew.com/article
/1994/12/30/entertainer-4-heather-locklear/.

24   **Women characters in media were more likely:** Dinitia Smith,
"Media More Likely to Show Women Talking About Romance
Than at a Job, Study Says," *New York Times*, May 1, 1997.

25   **"the most screwed-up Real Worlder ever":** Chris Hewitt
and Jim Walsh, "A New Reality: The Sixth Season of MTV's
Video-Verite 'The Real World' Is a Cause for Celebration among
Those Who Are Beyond Mere Fans," *St. Paul Pioneer Press*,
July 27, 1997, https://savesb6.newsbank.com:8443/MNGsave
/classic/doc?docid=1204002&q=(%20(video)%20)%20AND%20

date(07/27/1997%20TO%2007/27/1997)&stem=false&spa
ceop=AND&ttype=xsl&tval=headline_mng&pos=1&hn=
2&pubAbbrev=mng&dtokey=zouarqhlgykimtyiqunlrdww
#anchor1204002.

25 **volunteered with AIDS patients:** Matthew Scott Donnelly, "Tami Roman of 'Real World: Los Angeles' . . . Where Is She Now?" MTV News, December 19, 2011, http://www.mtv.com /news/2382184/where-is-real-world-tami-roman-now/.

26 **"Why did you kick David out":** Steve Weinstein, "A 'Real World' of Difference," *Los Angeles Times*, September 19, 1993.

26 **looking to cast "stereotypes":** Rick Jervis, "Real World, Few Vacancies: Hundreds Jockey for Spots When MTV Comes to SOBE," *Miami Herald*, November 18, 1995.

26 **"I could be the bitch":** Jervis, "Real World, Few Vacancies."

26 **"just like those live cop shows":** Sam Whiting, "Chance to Get Real on MTV Hit Serial S.F. Locale for New 'Real World,'" *San Francisco Chronicle*, October 18, 1993.

26 **"cute and sassy, but a two-faced back-stabber":** Hewitt and Walsh, "A New Reality."

26 **"bitch slap" was named:** Jennifer L. Pozner, *Reality Bites Back: The Troubling Truth About Guilty Pleasure TV* (Berkeley, CA: Seal Press, 2010).

26 **"what happens when people stop":** Nancy Sidewater, "If We Ran Reality TV; Paris, You're In. Trista, You're Out. We've Got an Extreme Makeover for Television's Most Popular—and Polarizing—Genre," *Entertainment Weekly*, May 21, 2004.

28 **Casting Doherty as a witch:** Ann Hodges, "Fall TV's Hits & Misses," *Houston Chronicle*, September 6, 1998.

28 **"a procession of attractive males":** Michael P. Lucas, "The Real Stories Behind a Trio of 'Charmed' Lives," *Los Angeles Times*, February 3, 1999.

28 **"*Charmed* is a perfect postfeminist girl-power show":** Bruce Fretts, "The Women of the WB Wow Audiences," *Entertainment Weekly*, December 25, 1998.

29 **"undead between dates":** Paula Geyh, "Feminism Fatale? 'Bad Girls' Adapt Women's Movement to Suit Themselves," *Chicago Tribune*, July 26, 1998.

CHAPTER 2: SEX IN THE 90s

32   **"Teachers say, 'Let's gross these kids out'":** Laura Sessions Stepp, "Beyond AIDS: Teenagers and STDs," *Washington Post,* March 23, 1999.

31   **$50 million per year:** Marcela Howell and Marilyn Keefe, "The History of Federal Abstinence-Only Funding," Advocates for Youth, July 2007, http://www.advocatesforyouth.org /publications/publications-a-z/429-the-history-of-federal -abstinence-only-funding.

33   **By 1994, the disease:** "A Timeline of HIV and AIDS," AIDS .gov, last updated 2016, accessed September 2016, https:// www.hiv.gov/hiv-basics/overview/history/hiv-and-aids -timeline.

33   **More than 440,000 cases:** Lawrence K. Altman, "AIDS Is Now the Leading Killer of Americans from 25 to 44," *New York Times,* January 31, 1995, https://partners.nytimes.com/library /national/science/aids/013195sci-aids.html.

33   **In 1996, women comprised 20 percent:** Centers for Disease Control and Prevention, "Update: Trends in AIDS Incidence, Deaths, and Prevalence—United States, 1996," *MMWR Weekly,* February 28, 1997, https://www.cdc.gov/mmwr/preview/mmwr html/00046531.htm.

33   **"European countries tend to be more open":** Debra W. Haffner, ed., *Facing Facts: The Report of the National Commission on Adolescent Sexual Health* (New York: National Commission on Adolescent Sexual Health, 1995).

34   **fifteen million Americans were contracting STDs:** "1998-MTV Sex in the 90's XII Fact or Fiction STD's-FULL EPISODE," MTV, YouTube video, aired March 1999, 21:50, posted by Matty Brown VHS Archives, August 2, 2015, https://www .youtube.com/watch?v=gI_WQYckPeQ.

34   **"Cat Daddies":** John Pope, "STD Rise in Teens Blamed on Older Men; 'Cat Daddies' Lure Girls, Infect Them, Officials Say," *New Orleans Times-Picayune,* August 15, 1999.

34   **it was collegiate women:** Author interviews with half a dozen women who were in college in the 90s.

35 **"There are no Magic Johnsons"**: Stepp, "Beyond AIDS: Teen-agers and STDs."

35 **"a detention-class film strip"**: Mike Flaherty, "Sex in the '90s XII: Fact or Fiction?" *Entertainment Weekly*, March 26, 1999, http://www.ew.com/article/1999/03/26/sex-90s-xii-fact-or-fiction.

35 **"Sex is famously hot hot hot"**: "1998-MTV Sex in the 90's XII Fact or Fiction STD's-FULL EPISODE," YouTube video.

36 **Jennie Miller was ashamed to discover**: Perri Peltz and Hugh Downs, "Intimate Danger," ABC News Network, September 3, 1999.

36 **An executive producer**: Children Now, "Reflections of Girls in the Media" (presentation, 4th Annual Children and the Media Conference, Los Angeles, April 30–May 2, 1997), http://files.eric.ed.gov/fulltext/ED433131.pdf.

38 *Seventeen, Teen,* **and** *YM* **circulation**: Diane Seo, "Magazines for Teens Thrive as Numbers, Buying Power Grow," *Los Angeles Times*, April 9, 1998.

38 **"Objectification theory" explained**: Barbara L. Fredrickson and Tomi-Ann Roberts, "Objectification Theory: Toward Understanding Women's Lived Experiences and Mental Health Risks," *Psychology of Women Quarterly* 21, no. 2 (1997): 173–206, http://www.sanchezlab.com/pdfs/FredricksonRoberts.pdf.

40 **In 1994, sociology professors**: Tamar Lewin, "Sex in America: Faithfulness in Marriage Thrives After All," *New York Times*, October 7, 1994, http://www.nytimes.com/1994/10/07/us/sex-in-america-faithfulness-in-marriage-thrives-after-all.html?pagewanted=2.

40 **"Whatever gauzy, Fabio-induced fantasies they had"**: Orenstein, *Flux.*

41 **When you consider that one in four women**: Nancy Gibbs, "When Is It Rape?" *Time*, June 3, 1991, http://content.time.com/time/subscriber/article/0,33009,973077-1,00.html.

41 **A Pentagon report later found**: Eric Schmitt, "Senior Navy Officers Suppressed Sex Investigation, Pentagon Says," *New York Times*, September 25, 1992, http://www.nytimes.com/1992/09/25/us/senior-navy-officers-suppressed-sex-investigation-pentagon-says.html.

42   **"If I could get every young person":** Claudia Dreifus, "Joy-celyn Elders," *New York Times*, January 30, 1994, http://www .nytimes.com/1994/01/30/magazine/joycelyn-elders.html?page wanted=all&mcubz=0.

42   **"is something that's part of human sexuality":** Patricia Smith, "Elders' Sin: Delivering Us from Ignorance," *Boston Globe,* December 14, 1994.

43   **"pardoned Pee-Wee Herman":** Salmagundi, "Hey, Kiddies, You Sure Were Fooled If You Fell for That . . ." *Baltimore Sun*, January 3, 1995, http://articles.baltimoresun.com/1995-01-03 /news/1995003168_1_joycelyn-elders-claus-nicholas.

43   **"filled with teenage boys":** Salmagundi, "Hey, Kiddies, You Sure Were Fooled if You Fell for That . . ."

43   **In a Top Ten list of least-convincing alibis:** "Late Show Top Ten," accessed November 8, 2017, http://www.oocities.org/jay lipp/Letterman/topten95.html.

44   **"went out of its way to make clear that her resignation had not been voluntary":** Douglas Jehl, "Surgeon General Forced to Resign by White House," *New York Times*, December 10, 1994, http://www.nytimes.com/1994/12/10/us/surgeon-general -forced-to-resign-by-white-house.html.

44   **"If she had not resigned":** Jehl, "Surgeon General Forced to Resign by White House."

44   **"I went to Washington feeling like prime steak":** "Then & Now: Joycelyn Elders," CNN, July 20, 2005, http://www.cnn .com/2005/US/07/18/cnn25.tan.elders/.

44   **forty thousand prescriptions:** Jacque Wilson, "Viagra: The Little Blue Pill That Could," CNN, March 27, 2013, http://www.cnn .com/2013/03/27/health/viagra-anniversary-timeline/index.html.

45   **"Being a horn dog was part of his persona."** Ty West, inter-view with the author, October 5, 2015.

45   **Monica Lewinsky herself had mocked:** Andrew Morton, *Monica's Story* (New York: St. Martin's Press, 1999), 123.

46   **"It was riveting to know":** Howard Kurtz, "Going Weak in the Knees for Clinton," *Washington Post*, July 6, 1998, http://www .washingtonpost.com/wp-srv/politics/special/clinton/stories /medianotes070698.htm.

46 **"He'd say, 'That's a great skirt'"**: Tabitha Soren, interview with the author, April 2015.

47 **a hand-wringing report**: Anne Jarrell, "The Face of Teenage Sex Grows Younger," *New York Times*, April 2, 2000, http://www.nytimes.com/2000/04/02/style/the-face-of-teenage-sex-grows-younger.html?pagewanted=all.

48 **"appealing to more than a small fraction of people"**: Lewin, "Sex in America: Faithfulness in Marriage Thrives After All."

48 **Researcher Laura Carpenter looked at two decades' worth**: Laura M. Carpenter, "From Girls into Women: Scripts for Sexuality and Romance in *Seventeen* Magazine, 1974–1994," *Journal of Sex Research* 35, no. 2 (May 1998).

CHAPTER 3: THE GOLDILOCKS CONUNDRUM

53 **"Perhaps this would be the dream job I had hoped for"**: Anita Hill, *Speaking Truth to Power* (New York: Anchor Books, 1998), 60.

53 **"I was extremely uncomfortable talking about sex"**: Anita Hill, "Opening Statement for the Senate Judiciary Committee," C-SPAN, YouTube video, aired October 11, 1991, 1:43:51, http://www.americanrhetoric.com/speeches/anitahillsenatejudiciarystatement.htm.

54 **This tactic dates back to slavery**: Cristen Conger, "The Legal History of Sexual Harassment," March 28, 2016, in *Stuff Mom Never Told You*, HowStuffWorks.com, podcast, 49:52, http://www.stuffmomnevertoldyou.com/podcasts/the-legal-history-of-sexual-harassment.htm.

54 **"It was immoral for young girls to be working alongside men"**: Tamar Lewin, "Archives of Business: Sexual Harassment in the Workplace; A Grueling Struggle for Equality," *New York Times*, November 9, 1986, http://www.nytimes.com/1986/11/09/business/archives-business-sexual-harassment-workplace-grueling-struggle-for-equality.html?pagewanted=all.

55 **Lin Farley coined it**: Lin Farley, "I Coined the Term 'Sexual Harassment.' Corporations Stole It," *New York Times*, October 18, 2017, https://www.nytimes.com/2017/10/18/opinion/sexual-harassment-corporations-steal.html.

55   **acknowledged the problem of sexual harassment:** Lewin, "Archives of Business: Sexual Harassment in the Workplace."

56   **"I certainly hope so":** Brooke Hauser, *Enter Helen: The Invention of Helen Gurley Brown and the Rise of the Modern Single Woman* (New York: HarperCollins, 2016).

56   **"unwelcome sexual advances":** "Sexual Harassment," US Equal Employment Opportunity Commission, accessed March 2016, http://www.eeoc.gov/laws/types/sexual_harassment.cfm.

57   **"I felt as though I had been dipped":** Hill, *Speaking Truth to Power,* 95.

59   **investigators were inexperienced with sexual harassment:** Melissa Harris-Perry, "EXCLUSIVE: Melissa Harris-Perry Interviews Anita Hill, 25 Years Later," *Essence,* March 30, 2016, http://www.essence.com/2016/04/06/exclusive-melissa-harris -perry-interviews-anita-hill-25-years-later.

59   **"I suspected that I would have been":** Hill, *Speaking Truth to Power,* 113.

59   **Hill's initial desire for anonymity:** Timothy Phelps and Helen Winternitz, *Capitol Games: Clarence Thomas, Anita Hill, and the Story of a Supreme Court Nomination* (New York: Hyperion, 1992), 231.

60   **"Is this the whole thing":** Phelps and Winternitz, *Capitol Games,* 240.

60   **"Whoever did this ought to be shot":** Phelps and Winternitz, *Capitol Games,* 234.

60   **Hill, her legal team:** Reuters, "THE THOMAS NOMINATION; Excerpts From Senate's Hearings on the Thomas Nomination," *New York Times,* October 12, 1991, http://www .nytimes.com/1991/10/12/us/the-thomas-nomination -excerpts-from-senate-s-hearings-on-the-thomas-nomination .html?pagewanted=1.

61   **"most embarrassing incidents alleged":** Committee on the Judiciary, *Nomination of Judge Clarence Thomas to Be Associate Justice of the Supreme Court of the United States* (Washington, DC: US Government Printing Office, 1993), https://www.loc.gov /law/find/nominations/thomas/hearing-pt4.pdf.

61   **"a foul stack of stench":** Judi Hasson, "A System Under Fire: Mud Bath Displaces Decorum," *USA Today,* October 14, 1991.

61 **"It was my opinion"**: Joe Battenfeld and Andrew Miga, "Hill Sticks to Her Story—Thomas Rips 'Sleaze,'" *Boston Herald*, October 12, 1991.

62 **"product of fantasy"**: Battenfeld and Miga, "Hill Sticks to Her Story."

62 **55 percent of men and 49 percent of women**: Priscilla Painton, "Woman Power," *Time*, October 28, 1991.

63 **Erotomania still appears**: Miles E. Drake Jr., "Delusional Disorder DSM-5 297.1 (F22)," *Theravive*, accessed July 2017, http://www.theravive.com/therapedia/Delusional-Disorder-DSM-5-297.1-(F22).

63 **"little is known about the background"**: N. Kennedy, M. McDonough, B. Kelly, and G. E. Berrios, "Erotomania Revisited: Clinical Course and Treatment," *Comparative Psychology* 43, no. 1 (January–February 2002): 1–6, https://www.ncbi.nlm.nih.gov/pubmed/11788912.

63 **"tactics against sexual harassment"**: Greg Scott and Brian Martin, "Tactics Against Sexual Harassment: The Role of Backfire," *Journal of International Women's Studies* 7, no. 4 (May 2006): 111–125, http://vc.bridgew.edu/cgi/viewcontent.cgi?article=1464&context=jiws.

63 **"a rare delusion of some women"**: John C. Danforth, *Resurrection: The Confirmation of Clarence Thomas* (New York: Viking, 1994), 155.

64 **"The issue of what I was doing"**: Dominick Dunne, "The Verdict," *Vanity Fair*, March 1992.

64 **"aggressive perpetrator"**: David Margolick, "Smith Tells Rapt Courtroom His Side of Story," *New York Times*, December 11, 1991, http://www.nytimes.com/1991/12/11/us/smith-tells-rapt-courtroom-his-side-of-story.html?pagewanted=all.

65 **"A stalker who tried to blackmail"**: Morton, *Monica's Story*, 203.

66 **"Why should I allow"**: Hill, *Speaking Truth to Power*, 138.

66 **"If what you say this man said to you occurred"**: Hugh Downs and Barbara Walters, "The Thomas Hearings: How Far Is Too Far?" *20/20* October 11, 1991.

66 **He presented these obscurities**: Ann McDaniel, "The Attack of the Bush Men," *Newsweek*, October 28, 1991.

66 **"They thought the more they pressed"**: Anita Hill, *Anita: Speaking Truth to Power*, directed by Freida Lee Mock (New York: First Run Features, 2014).

67 **"X-rated Drama in a Regal Setting"**: Andrew Miga, "X-rated Drama in a Regal Setting," *Boston Herald*, October 12, 1991.

67 **"the most lurid and dispiriting proceeding"**: Tom Shales, "Looking Back, Looking Ahead; Television '91: Onward & Downward; When the Lowest Common Denominator Fell Lower," *Washington Post*, December 29, 1991.

67 **"run from the TV set directly to the showers"**: Tom Shales, "The Hard-to-Watch Drama That Left No One Unscathed," *Washington Post*, October 12, 1991.

67 **"It's the first time we've got a story"**: Patrick Brogan, "Sexual Politics: Clarence Thomas—Victim of Lynch Mob or Lecher Exposed?" *Observer*, October 13, 1991.

68 **"impressed by the way Thomas"**: McDaniel, "The Attack of the Bush Men."

68 **"pretending to be offended"**: Jill Niebrugge-Brantley, "A Feminist Writes about the Anita Hill–Clarence Thomas Conflict," http://chnm.gmu.edu/courses/122/hill/brantley.htm.

68 **"African American Women in Defense of Ourselves"**: Elsa Barkley Brown, Deborah King, and Barbara Ransby, "African American Women in Defense of Ourselves," *New York Times*, November 17, 1991.

68 **Hill was a victim of intersectional disempowerment**: Kimberlé Williams Crenshaw, "Black Women Still in Defense of Ourselves," *Nation*, October 5, 2011, https://www.thenation.com/article/black-women-still-defense-ourselves/.

69 **another black female associate:** Phelps and Winternitz, *Capitol Games*, 398.

70 **"I knew that Clarence Thomas was capable"**: Michel Martin, "Justice Clarence Thomas Accuser Angela Wright Says Thomas 'Perjured Himself onto the Supreme Court,'" *NPR News*, October 9, 2007, http://www.npr.org/about/press/2007/100907.wright.html.

70 **"If you were young"**: Ruth Marcus, "One Angry Man," *Washington Post*, October 3, 2007, http://www.washingtonpost.com

/wp-dyn/content/article/2007/10/02/AR2007100201822
.html.

70 **On her first day at work:** Morton, *Monica's Story*, 62.

71 **It spanned more than two years:** Morton, *Monica's Story*, 5.

71 **"immature and inappropriate behavior":** "A Chronology: Key Moments in the Clinton-Lewinsky Saga," CNN, 1998, http:// www.cnn.com/ALLPOLITICS/1998/resources/lewinsky /timeline/.

71 **Lewinsky was devastated:** Morton, *Monica's Story*.

71 **Tripp surreptitiously recorded:** "A Chronology: Key Moments in the Clinton-Lewinsky Saga," CNN.

72 **On January 12, Tripp called:** Morton, *Monica's Story*, 170.

72 **"the most riveting chapter of recent American history":** Richard A. Posner, "An Affair of State: The Investigation, Impeachment, and Trial of President Clinton," *New York Times*, 1999, https://www.nytimes.com/books/first/p/posner-affair .html.

73 **"low-cut blouses" and "thigh-high skirts":** Laura Ingraham, "'All About Monica'—Lessons From a Morality Play," *Los Angeles Times*, May 3, 1998, http://articles.latimes.com/1998 /may/03/opinion/op-45872.

73 **When she visited her father in Los Angeles:** Michael Isikoff and Evan Thomas, "The Eye of the Storm," *Newsweek*, February 16, 1998.

74 **"Clinton bimbo":** Dorothy Rabinowitz, "Juanita Broaddrick Meets the Press," *Wall Street Journal*, February 19, 1999, http:// www.anusha.com/wsjbroad.htm.

74 **"Big Mac":** Morton, *Monica's Story*, 23.

74 **The repeated fat-shaming stung:** Morton, *Monica's Story*, 62.

74 **They reported on what they thought she ate:** Vanessa Grigoriadis, "Monica Takes Manhattan," *New York*, March 19, 2001, http://nymag.com/nymetro/news/people/features/4481/.

75 **"overlit her face to make it appear thinner":** Bill Hoffman, "The Sultry Pepperpot!; Di's Fotog Brings Out Best in Monica," *New York Post*, March 19, 1999, http://nypost.com/1999/03/19 /the-sultry-pepperpot-dis-fotog-brings-out-best-in-monica/.

75 **"the girl who was too tubby":** Maureen Dowd, "President

Irresistible," *New York Times*, February 18, 1998, http://www
.nytimes.com/1998/02/18/opinion/liberties-president-irresistible
.html.

75   **"Before she became obsessed":** Jeffrey Toobin, *A Vast Conspiracy: The Real Story of the Sex Scandal That Nearly Brought Down a
President* (New York: Simon & Schuster, 1999), 84.

75   **"chubby" and "cute, if a little zaftig":** Jake Tapper, "I Dated
Monica Lewinsky," *Washington City Paper*, January 30, 1998,
http://www.washingtoncitypaper.com/articles/14334/i
-dated-monica-lewinsky.

75   **"as an oversized":** Jennifer-Scott Mobley, "Fatsploitation: Dis-
gust and the Performance of Weight Loss," in *Fat: Culture and
Materiality*, eds. Christopher E. Forth and Alison Leitch (New
York: Bloomsbury Academic, 2014).

76   **"for her fresh and insightful columns":** "1999 Pulitzer Prizes
for Journalism," Pulitzer Prizes, accessed November 4, 2017,
http://www.pulitzer.org/prize-winners-by-year/1999.

77   **"ditsy, predatory White House intern":** Amanda Hess,
"'Ditsy, Predatory White House Intern,'" *Slate*, May 7, 2014,
http://www.slate.com/articles/double_x/doublex/2014/05
/monica_lewinsky_returns_how_maureen_dowd_caricatured
_bill_clinton_s_mistress.html.

77   **"the Troubled Slut Defense":** Dowd, "President Irresistible."

77   **"Moremean Dowdy":** Monica Lewinsky, "Exclusive: Monica
Lewinsky Writes about Her Affair with President Clinton," *Vanity
Fair*, May 6, 2014, http://www.vanityfair.com/news/2014/05
/monica-lewinsky-speaks.

77   **In the August 1998 column:** Maureen Dowd, "Monica Gets
Her Man," *New York Times*, August 23, 1998, http://www
.nytimes.com/1998/08/23/opinion/liberties-monica-gets
-her-man.html.

77   **He remarked to her father:** Morton, *Monica's Story*, 229.

78   **Monica Lewinsky was accused of suffering:** Maureen Dowd,
Are Men Necessary? (New York: Penguin, 2005), 81.

79   **It found that women were:** Patricia Tjaden and Nancy
Thoennes, "Stalking in America: Findings from the National
Violence against Women Survey," National Institute of Justice

Centers for Disease Control and Prevention, April 1998, https://www.ncjrs.gov/pdffiles/169592.pdf.

80 **"the most-watched news interview":** Brian Lowry and Elizabeth Jensen, "Huge Ratings for Lewinsky," *Los Angeles Times*, March 5, 1999, http://articles.latimes.com/1999/mar/05/entertainment/ca-14057.

80 **Glaze by Club Monaco:** Barbara Walters, *Audition: A Memoir* (New York: Knopf Doubleday, 2008).

80 **"People felt free to leave the most cruel and revolting messages imaginable":** Hill, *Speaking Truth to Power*, 129.

80 **"We felt that people would focus on her and not stay with the institutional message":** "Liz and Larry in Morocco. Notes from All Over on Hill and Thomas," *Toronto Star*, October 27, 1991.

81 **"fatal assistant":** Virginia Lamp Thomas, "Breaking Silence," *People*, November 11, 1991, http://people.com/archive/cover-story-breaking-silence-vol-36-no-18/.

81 **Hill had myomectomy surgery:** Hill, *Speaking Truth to Power*, 261.

81 **"effectively committed perjury":** David Brock, *Booknotes*, C-SPAN, June 13, 1993, http://www.booknotes.org/Watch/43009-1/David-Brock.

83 **The all-male Senate Judiciary Committee:** Caryl Rivers, "Perspectives on Testimony; Is There No Believable Woman?" *Los Angeles Times*, December 12, 1991.

83 **biblical women:** *Anita: Speaking Truth to Power*, directed by Mock.

83 **"less about success than survival":** Hill, *Speaking Truth to Power*, 99.

83 **In the six years following:** Nina Totenberg, "Thomas Confirmation Hearings Had Ripple Effect," NPR, October 11, 2011, http://www.npr.org/2011/10/11/141213260/thomas-confirmation-hearings-had-ripple-effect.

84 **"Being consumed with anger":** Leslie Bennetts, "Anita Hill Discusses Clarence Thomas Twenty Years Later," *Newsweek*, October 3, 2011, http://www.newsweek.com/anita-hill-discusses-clarence-thomas-twenty-years-later-68241.

85 **a reason the kind of harassment she experienced persists**

today: Gretchen Carlson, conversation at TEDWomen, November 2, 2017.

85 **More than 70 percent of adults:** Maeve Duggan, "Online Harassment," Pew Research Center, October 22, 2014, http://www.pewinternet.org/2014/10/22/online-harassment.

85 **"Where is the guy brave enough":** Richard Cohen, "Fairness for Lewinsky," *Washington Post*, January 2, 2007, http://www.washingtonpost.com/wp-dyn/content/article/2007/01/01/AR2007010100701.html.

86 **reportedly using tax dollars to do so:** Jim Ruttenberg and Robbie Brown, "Governor Used State's Money to Visit Lover," *New York Times*, June 25, 2009, http://www.nytimes.com/2009/06/26/us/26sanford.html.

87 **Through tears, she thanked her own tormentors:** Alexandra Schwartz, "Monica Lewinsky and the Shame Game," *New Yorker*, March 26, 2015, http://www.newyorker.com/culture/cultural-comment/monica-lewinsky-and-the-shame-game.

CHAPTER 4: WOMEN WHO WORKED

90 **"What was it that I did personally to make this happen?":** Marcia Clark, interview with the author, November 6, 2015.

90 **"Nobody has a mind like Marcia":** Lorraine Adams, "The Fight of Her Life; Marcia Clark—Working Mother and O. J. Simpson's Lead Prosecutor—Takes Her Place Among Other Maligned, Adored and Misunderstood Modern Women," *Washington Post*, August 20, 1995.

91 **"And I remember saying to people":** Marcia Clark, interview by Cynthia McFadden, 92nd Street Y Talks, May 26, 2016, http://92yondemand.org/marcia-clark-conversation-cynthia-mcfadden.

91 **"Trial work is especially appealing":** Marcia Clark, *Without a Doubt* (New York: Penguin Books, 1998), 7.

92 **has also studied this phenomenon:** Amy Cuddy, interview by Ben Lillie, "In Debates, Watch for Signs of Warmth," TED Blog, October 1, 2012, http://blog.ted.com/in-debates-watch-for-signs-of-warmth-qa-with-amy-cuddy/.

92   **Critics called her ruthless:** Gerard Evans, "A Tigress Who Is Trying to Soften Her Claws," *Daily Mail,* January 27, 1995.

93   **During jury selection:** Linda Deutsch, "Jury Prospect: O. J. Simpson is 'A Hunk,' Clark's Skirts Too Short," Associated Press, October 28, 1994.

93   **They were "her one vanity":** Gerard Evans, "L.A CLAWS!; Glamour Girl Battle over OJ; Prosecutor Marcia Clark Who Is Involved in OJ Simpson Trial," *Daily Record,* September 28, 1994.

93   **Trial watchers were subject to:** Richard Price and Haya El Nasser, "If Looks Could Convict . . . / Prosecutor Softens Up Her Style," *USA Today,* October 10, 1994.

93   **She was "an attractive lady":** Jeffrey Toobin, "True Grit," *New Yorker,* January 9, 1995.

93   **After a prospective juror:** Deutsch, "Jury Prospect: O. J. Simpson Is 'A Hunk'; Clark's Skirts Too Short."

93   **"privately expressed":** Evans, "L.A CLAWS!; Glamour Girl Battle Over OJ."

94   **"If it's too short":** Francine Parnes, "Court Appearances Count," Associated Press, August 12, 1994, accessed December 11, 2017.

94   **The country also fixated:** Susan Reimer, "Marcia Clark's Trials Have Now Begun Outside the Courtroom," *Baltimore Sun,* March 5, 1995, http://articles.baltimoresun.com/1995-03-05/features/1995064145_1_marcia-clark-prosecutor-marcia-simpson-trial.

94   **"most memorable mole":** Daniel Margolick, "The Murder Case of a Lifetime Gets a Murder Prosecutor of Distinction," *New York Times,* January 22, 1995.

94   **"more than her share of bad hair days":** Margaret Wente, "The Adventures of Marcia Cark," *Globe and Mail* (Canada), March 18, 1995.

94   **"poodle 'do":** Teresa Wiltz, "Some Fashion Crimes in the Simpson Courtroom," *Chicago Tribune,* January 26, 1995.

94   **"a mole painted just above":** Salley McInerney, "A Haunting Halloween," *Atlanta Journal-Constitution,* October 31, 1995.

95   **"Heather was very hot then":** Diane English, interview with the author, February 17, 2016.

95   **Since she contractually couldn't edit:** Meredith Blake, "25 Years Later, Looking Back at 'Murphy Brown,'" *Los Angeles Times*, December 23, 2013, http://articles.latimes.com/2013/dec/23/entertainment/la-et-st-murphy-brown-20131223.

96   **"bitchy but lovable":** Norm Schaefer, "It's Prime Time for Women on Networks," *Chicago Sun-Times*, September 11, 1994.

96   **"finally catching on to the reality":** Schaefer, "It's Prime Time for Women on Networks."

97   **In other words:** "NOW Report—Women on TV," United Press International, November 3, 2002, https://www.upi.com/Archives/2002/11/03/Feature-NOW-report-women-on-TV/1771036299600/.

97   **After winning her fifth in 1995:** "Murphy Brown Trivia," IMDb, accessed November 4, 2017, http://www.imdb.com/title/tt0094514/trivia.

97   **Since its inception, television network Fox:** "Black Is Bountiful," *Newsweek*, December 5, 1993.

98   **Soon, the show beat:** Greg Braxton, "'Living Single' Is Living Large on FOX: Despite Criticism over Male-Bashing and Sexual References, the Show Has Found an Audience," *Los Angeles Times*, December 9, 1993, http://articles.latimes.com/1993-12-09/entertainment/ca-118_1_young-black-men.

98   **"a bunch of fat, happy women":** Braxton, "'Living Single' Is Living Large on FOX."

99   **"is predicated on the presence":** K. S. Jewell, *From Mammy to Miss America and Beyond: Cultural Images and the Shaping of US Social Policy* (New York: Routledge, 1993).

99   **Critics' arguments that:** Greg Braxton, "'Living Single' Is Living Large on FOX."

99   **The black women on *Living Single* were stereotyped:** "Black Is Bountiful," *Newsweek*, December 5, 1993, http://www.newsweek.com/black-bountiful-190658.

99   **"quadruple the sex drive":** "Black is Bountiful," *Newsweek*.

99   **"oversexed, wha's-up, man buffoons":** Braxton, "'Living Single' Is Living Large on FOX."

99   **Bill Cosby lambasted the show:** Bill Cosby, "Someone at the Top Has to Say: 'Enough of This,'" *Newsweek*, December 6, 1993.

100 **"You're not going to get your female show"**: James Stern-gold, "Strong Women In TV? They'd Sure Better Be; Progress, But Slowly, On Camera And Off," *New York Times*, December 18, 1997, http://www.nytimes.com/1997/12/18/arts/strong-women -in-tv-they-d-sure-better-be-progress-but-slowly-on-camera -and-off.html.

101 **While the *Friends* cast would:** Greg Braxton, "'Single' Looks for a Little Help Against 'Friends,'" *Los Angeles Times*, February 5, 1996, http://articles.latimes.com/1996-02-01/entertain ment/ca-30941_1_single-friends-living.

101 **Between 1990 and 1999:** *Women in America*, US Census Bureau.

102 **"beribboned husband-catching primer":** Leisl Schillinger, "Desperately Seeking Simon," *The Independent*, June 1, 1996, http://www.independent.co.uk/life-style/desperately-seeking -simon-1335028.html.

103 **breast augmentations increased more than 700 percent:** Ariel Levy, *Female Chauvinist Pigs: Women and the Rise of Raunch Culture* (New York: Free Press, 2005), 22.

103 **"Women want to get married":** "Dateline: The Rules," You-Tube video, 7:03, posted by The Rules Book Official Channel, February 20, 2015, https://www.youtube.com/watch?v= W_xvzuQA1b4.

103 **"forget equality":** Laurette Ziemer, "It's the Dating Game; Play Hard to Get to Catch Your Mr. Right," *Evening Standard*, May 14, 1996.

103 **"another manifestation of the New Conservatism":** Diane White, "Man Hunting? Listen to Mom," *Boston Globe*, March 9, 1995.

104 **"Thousands of unsuspecting American males":** Ziemer, "It's the Dating Game."

104 **"boot camp . . . with the drill sergeants":** "Dateline: The Rules," YouTube video.

104 **"If you have sex with him too soon":** Bradley Gerstman, Christopher Pizzo, and Rich Seldes, *What Men Want: Three Professional Single Men Reveal to Women What It Takes to Make a Man Yours* (New York: HarperCollins, 1998), 18.

105 **roused the alpha male seduction communities:** Meg

Barker, "Rewriting the rules: Dr. Meg Barker at TEDxBrighton," TEDxBrighton, YouTube video, 11:41, posted by TEDx Talks, December 2, 2013, https://www.youtube.com/watch?v=XUOQprqrxFg.

105 **Matt Lauer interviewed:** *Today,* https://www.youtube.com/watch?v=ZbU8kc9WFrQ.

106 ***Mars/Venus* had sold six million copies:** Elizabeth Gleick, "Tower of Psychobabble," *Time,* June 16, 1997.

107 **"looking peacockish":** Adams, "The Fight of Her Life."

107 **"She was just flying around":** David Bowman, "Profiler: John Douglas," *Salon,* June 8, 1999, http://www.salon.com/1999/07/08/profiler/.

108 **"Only Hillary Clinton has gone through more":** Reimer, "Marcia Clark's Trials Have Now Begun Outside the Courtroom."

110 **"For the next month, he":** Marcia Clark interview with the author, May 2015.

111 **"the greatest irony of all":** Morton, *Monica's Story,* 81.

112 **"She uses all of her resources":** Alan Dershowitz interview by Martin Savidge, *New Day,* CNN, November 17, 2013.

113 **"is a projection of men's preoccupation with sex":** Alessandra Stanley, "Erotomania: A Rare Disorder Runs Riot—in Men's Minds," *New York Times,* November 10, 1991.

113 **"a flirt in lawyer's clothing":** "Greta, You're One Sharp Dresser," *Roanoke Times,* April 8, 1996.

116 **"I like him very much":** Sarah Paulson, interview by Terry Gross, *Fresh Air,* NPR, March 10, 2016, http://www.npr.org/2016/03/10/469922588/sarah-paulson-strives-to-get-it-right-as-o-j-simpson-prosecutor.

116 **"sexless, old and a bitch":** Ruth Richman, "Getting Real: TV Slowly Coming into Focus on Women in the Workplace," *Chicago Tribune,* May 18, 1997.

117 **"She is young, perennially confused":** Nancy Hass, "Hard Times for Strong-Minded Women," *New York Times,* September 27, 1998, http://www.nytimes.com/1998/09/27/arts/television-hard-times-for-strong-minded-women.html.

117 **"dizzy girl-woman"**: Martha M. Lauzen, "Alpha Females Still Trail Adorable Dopes," *Los Angeles Times*, December 13, 1999, http://articles.latimes.com/1999/dec/13/entertainment /ca-43357.

117 **"scarfing down low-fat Doritos"**: Hass, "Hard Times for Strong-Minded Women."

118 **pioneered a certain type of third-wave feminist antihero**: Emily Nussbaum, "Difficult Women," *New Yorker*, July 29, 2013, http://www.newyorker.com/magazine/2013/07/29/difficult -women.

118 **"Her indecision and compelling need to please others"**: Lauzen, "Alpha Females Still Trail Adorable Dopes."

118 **"simply drop off the primetime planet," according to Lauzen**: Tom Tapp, "New Survey Says Tube Not Tops for Women," *Daily Variety*, September 17, 1999.

119 **"I plan to change it"**: Orenstein, *Flux*, 3.

119 **"emotional klutziness of a teenager"**: Lynn Elber, "Women Power Has Become Girl Power on the Tube," *Chicago Tribune*, June 15, 1998.

119 **"regularly defeated by her neuroses"**: Joanne Ostrow, "Smart Chicks: A New Kind of Heroine Emerges as TV Role Model," *Denver Post*, September 30, 2004.

119 **"wimp"**: Jennifer L. Pozner, "And the Category Is . . . Simpering Wimps for $1,000,'" *Sojourner*, September 1998.

120 **"frustrated by feminist arguments"**: Joanne Ostrow, "Writer Kelley Presents Keen Understanding of Women," *Denver Post*, January 15, 1998.

120 **"bra burning to ohmigosh, I just wanna have sex!"**: Kathleen Parker, "Feminism Isn't Dead, Just Bored and Confused," *Orlando Sentinel*, June 27, 1998.

121 **"The show's shoe fetish has given way to more Fendi bags"**: Mimi Avins, "Heights of Fame: She May Not Be as Flashy as Her Character on 'Sex and the City,' but Sarah Jessica Parker Still Flirts with the Outrageous," *Los Angeles Times*, October 3, 1999.

121 **"For all the frantic coupling"**: Wendy Shalit, "Lonely Liberation," *Washington Times*, October 27, 1999.

CHAPTER 5: BAD MOM

124 **"I have informed the court that I cannot be present"**: Bettina Boxall, "Marcia Clark's Husband Says She Misled Ito," *New Orleans Times-Picayune*, March 4, 1995.

124 **"She has no childcare problem"**: "You Have to Care for the Kids," *Newsweek*, April 17, 1995.

124 **"While I commend her brilliance"**: Wente, "The Adventures of Marcia Clark."

124 **"I have personal knowledge that on most nights"**: Clarence Page, "Who's Listening to the Kids in Custody Battles?" *Chicago Tribune*, March 12, 1995.

124 **"putting a wife to work"**: David Wright, "Trump in 1994: 'Putting a Wife to Work Is a Very Dangerous Thing,'" CNN, June 2, 2016, http://www.cnn.com/2016/06/02/politics/trump -wife-comments-abc-interview/index.html.

125 **"winning any Good Mother awards this year"**: Margery Eagan, "Respect for Motherhood a Casualty of Our Times," *Boston Herald*, March 9, 1995.

125 **"'Nurturing' is not a word in her lexicon"**: Wente, "The Adventures of Marcia Clark."

125 **A lawyer in the office**: Peter S. Canellos, "Clark Case Sparks Debate on Work and Gender Roles," *Boston Globe*, March 4, 1995.

125 **a Michigan woman lost custody**: "Baby Maranda Case: 1994," *Law Library*, http://law.jrank.org/pages/3612/Baby-Maranda -Case-1994.html.

126 **a child was taken away from a mother**: Susan Chira, "Custody Case Stirs Debate on Bias Against Working Women," *New York Times*, July 31, 1994, http://www.nytimes.com/1994/07/31/us /custody-case-stirs-debate-on-bias-against-working-women.html ?pagewanted=all.

126 **"lingering bias that penalizes mothers"**: Chira, "Custody Case Stirs Debate on Bias Against Working Women."

126 **"dump a dad, get a check"**: Anne P. Mitchell and Wolfgang Hirczy, "Happy Fatherless Day; Are We Sending a Message That Dads Are Disposable?" *Washington Post*, June 18, 1995.

127 **half of Massachusetts probate judges:** Chira, "Custody Case Stirs Debate on Bias Against Working Women."

129 **"a little like Monticello":** Karl Vick, "Baird Gets Opportunity to Give Her Side of the Story," *St. Petersburg Times*, January 20, 1993.

129 **"servants" to maintain a household:** Robert Kuttner, ". . . And Double Standards," *Washington Post*, January 22, 1993.

129 **"Widespread anger":** Victoria Benning and Bob Hohler, "Women's Groups Look Past Baird; Some Think Excessive Response But Voice Preference for Other Candidates," *Boston Globe*, January 22, 1993.

130 **"There are . . . millions of Americans out there":** John Farrell Aloysius, "Baird Apologizes in Illegal-Alien Hiring," *Boston Globe*, January 20, 1993.

131 **"The reaction to Baird's admission":** Benning and Hohler, "Women's Groups Look Past Baird."

132 **NOW's six hundred affiliates:** *World News Tonight with Peter Jennings*, ABC News, February 8, 1993, accessed December 19, 2017.

132 **Ronald Brown revealed:** Phil Mintz, "Trouble at Home; Brown Admits He Owed Tax for Cleaner," *Newsday*, February 8, 1993.

132 **"declined to even rebuke a male cabinet member":** Susan Page, "Clinton on Defensive; Under Attack on Cabinet Jobs," *Newsday*, February 9, 1993.

132 **When Labor Secretary:** "Commerce Secretary Ron Brown Facing Questions about Domestic Help," CBS News transcripts, *CBS This Morning*, February 8, 1993.

133 **"There's no doubt":** Howard Fineman, Mark Miller, and Joe Klein, "Hillary Clinton, First Lady: The 1993 Newsweek Cover Story," *Newsweek*, April 11, 2015, http://www.newsweek.com /hillary-clinton-first-lady-newsweeks-1993-cover-story-32 1514.

133 **"television's first feminist and working-class-family sitcom":** Roseanne Barr, "And I Should Know," *New York*, May 15, 2011, http://nymag.com/arts/tv/upfronts/2011/rose anne-barr-2011-5/.

133 **she broke a record for appearing:** David Plotz, "Domestic

Goddess Dethroned: How Roseanne Lost It," *Slate*, May 18, 1997, http://www.slate.com/articles/news_and_politics/assessment/1997/05/domestic_goddess_dethroned.html.

134 **"I think women should be more violent"**: Corky Siemaszko, "She Has a Rosie View of Violence," *New York Daily News*, July 10, 1995.

134 **"Her raucous antics"**: John J. O'Connor, "By Any Name, Roseanne Is Roseanne Is Roseanne," *New York Times*, August 18, 1991, http://www.nytimes.com/1991/08/18/arts/tv-view-by-any-name-roseanne-is-roseanne-is-roseanne.html?pagewanted=all.

135 **skipped work to chug**: "Welfare Queens," September 7, 2016, in *Stuff Mom Never Told You*, HowStuffWorks.com, podcast, 1:10:47, http://www.stuffmomnevertoldyou.com/podcasts/welfare-queens.htm.

135 **"narrative script"**: Franklin D. Gilliam Jr., "The 'Welfare Queen' Experiment," *Nieman Reports*, June 15, 1999, http://niemanreports.org/articles/the-welfare-queen-experiment/.

135 **The news media in the 90s**: Gilliam, "The 'Welfare Queen' Experiment."

135 **Such portrayals stoked America's belief**: Gilliam, "The 'Welfare Queen' Experiment."

136 **"reduce non marital births and encourage marriage"**: Rebecca Blank, "Evaluating Welfare Reform in the United States," *Journal of Economic Literature* 40, no. 4 (2002).

136 **"one of the most bitter chapters"**: Blake, "25 Years Later, Looking Back at 'Murphy Brown.'"

136 **baseball's World Series final**: Wikipedia, s.v. "World Series television ratings," last modified June 14, 2017, https://en.wikipedia.org/wiki/World_Series_television_ratings.

136 **"implying that she had been unnatural before"**: Susan J. Douglas, *The Rise of Enlightened Sexism: How Pop Culture Took Us from Girl Power to Girls Gone Wild* (New York: St. Martin's Press, 2010), 40.

137 **"deviant behavior"**: Barbara Dafoe Whitehead, "Dan Quayle Was Right," *Atlantic*, April 1993.

137 **"restoration of family values"**: Kenneth D. Whitehead, "Family

Values, Moral Values," *Catholic League Newsletter*, July 20, 1993, http://www.catholicleague.org/family-values-moral-values/.

137 **"devastating . . . as even Murphy Brown would admit"**: Joan Beck, "A Father's Ultimate Betrayal," *Chicago Tribune*, March 2, 1995, http://articles.chicagotribune.com/1995-03-02/news/9503020076_1_fatherless-families-fatherless-children-fatherless-america.

138 **"a weapon in the right's attack on single motherhood"**: James Atlas, "The Counter Counterculture," *New York Times*, February 12, 1995, http://www.nytimes.com/1995/02/12/magazine/the-counter-counterculture.html.

138 **"full-time mother of two"**: Candice Bergen, "Letter to the Editor: 'Murphy Brown's Values,'" *New York Times*, June 23, 1998, http://www.nytimes.com/1998/06/23/opinion/l-murphy-brown-s-values-543110.html.

139 **battling a fictional character**: Greg Braxton and John M. Broder, "It's Murphy Brown's Turn to Lecture Vice President: Television: Show Mixes Fiction and Reality as the Candice Bergen Character Responds to Quayle on Family Values Issue," *Los Angeles Times*, September 22, 1992, http://articles.latimes.com/1992-09-22/news/mn-1025_1_family-values.

140 **a two-year survey**: Center for the Advancement of Women, "Is Your Mother's Feminism Dead? New Agenda for Women Revealed in Landmark Two-Year Study; Violence and Equal Pay Top Women's Priorities; Followed by Health Care and Child Care; Abortion Rights Low on the List," *PR Newswire*, June 24, 2003.

CHAPTER 6: FIRST BITCH

142 **"This is supposed to be the year of the women in the Senate"**: George Bush, "Presidential Debate at the University of Richmond" (transcript), *American Presidency Project*, October 15, 1992.

142 **"Calling 1992 the 'year of the woman'"**: Emma Green, "A Lot Has Changed in Congress Since 1992, the 'Year of the Woman,'" *Atlantic*, September 26, 2013, http://www.theatlantic.com

/politics/archive/2013/09/a-lot-has-changed-in-congress-since
-1992-the-year-of-the-woman/280046/.

144 **"black-cent"**: Tim Dowling, "The Mystery of Hillary Clinton's
Changing Accent," *The Guardian*, May 1, 2007, http://www
.theguardian.com/world/2007/may/02/hillaryclinton
.uselections2008.

144 **"Lady Macbeth framing"**: Caroline Ervin and Cristen Conger,
"Who Is Hillary Clinton?" May 18, 2016, in *Stuff Mom Never Told
You*, HowStuffWorks.com, podcast, 1:26:27, https://www.stuff
momnevertoldyou.com/podcasts/who-is-hillary-clinton.htm.

145 **"the meek, mild, wronged wife"**: Malu Halasa, "A Law Unto
Herself," *The Guardian*, February 19, 1992.

145 **"knowingly lied about her husband's uncontainable sex
life"**: Christopher Hitchens, "The Case Against Hillary Clinton,"
Slate, January 14, 2008, http://www.slate.com/articles/news
_and_politics/fighting_words/2008/01/the_case_against
_hillary_clinton.html.

145 **"the longest, slowest, most painful car crash"**: Jake Tapper,
"The Clinton Marriage," *Salon*, August 26, 1999, http://www
.salon.com/1999/08/26/clintons/.

145 **"appearing to show contempt for women who work at
home"**: William Safire, "The Hillary Problem," *New York Times*,
March 26, 1992, http://www.nytimes.com/1992/03/26/opinion
/essay-the-hillary-problem.html.

145 **"everybody's grandmother"**: Donnie Radcliffe, "First Lady
Gets the Third Degree; On the Stump, Barbara Warms Them
Up. But George Burns Them Up," *Washington Post*, February 7,
1992.

145 **"mad as hell"**: Eric Snider, "Wynette Gets 'Justified,'" *St. Petersburg Times*, March 14, 1992.

145 **"I can assure you, in spite of your education"**: "Singer Tammy
Wynette Furious at Hillary Clinton's Statement," Associated
Press, January 29, 1992.

146 **"wild-eyed feminist who equated marriage with slavery"**:
Patricia McLaughlin, "Women of the House over the Years,
First Ladies Have Evolved into Capital Women," *Chicago Tri-
bune*, December 30, 1992.

146 **"Lips pulled back over her slightly jutting teeth"**: Gail Sheehy, "What Hillary Wants," *Vanity Fair*, May 1992, http://www.vanityfair.com/news/1992/05/hillary-clinton-first-lady-presidency.

147 **"Nothing too Hillary"**: Maureen Dowd, "Candidate's Wife; Hillary Clinton as Aspiring First Lady: Role Model, or a 'Hall Monitor' Type?" *New York Times*, May 18, 1992.

148 **"I want maneuverability"**: Dowd, "Candidate's Wife; Hillary Clinton as Aspiring First Lady."

148 **poll had found that:** Sheehy, "What Hillary Wants."

148 **"Would she work on the outside or the inside"**: Safire, "The Hillary Problem."

148 **"wants a First Lady to be an adjunct"**: Ted Koppel, "Making Hillary Clinton an Issue," *Nightline* transcript, PBS, March 26, 1992, http://www.pbs.org/wgbh/pages/frontline/shows/clinton/etc/03261992.html.

149 **"uncharacteristic"**: Wolf Blitzer, "Final Clinton Cabinet Appointments Followed by Singing," CNN, December 24, 1992.

149 **"rich, rich, rich chocolate cake"**: Marian Burros, "Bill Clinton and Food: Jack Sprat He's Not," *New York Times*, December 23, 1992, http://www.nytimes.com/1992/12/23/garden/bill-clinton-and-food-jack-sprat-he-s-not.html?pagewanted=all&mcubz=0.

149 **"She was, essentially, fired"**: Gil Troy, *The Age of Clinton: America in the 90s* (New York: Macmillan, 2015).

150 **"welcoming men to their role as the second sex"**: Christopher Caldwell, "The Feminization of America," *The Weekly Standard*, December 22, 1996, http://www.weeklystandard.com/the-feminization-of-america/article/9604.

150 **"deeply American fear of the unaccountable power behind the throne"**: Gil Troy, interview with the author during the 2012 presidential campaign.

151 **"hall monitor" whose "offputting" "drive and earnestness"**: Dowd, "Candidate's Wife; Hillary Clinton as Aspiring First Lady."

151 **"nattering cheerily"**: Alex Beam, "Warm Fuzzies from the First Lady," *Boston Globe*, January 3, 1996.

151 **"sublimated her into a nightmarish amalgam"**: Beam, "Warm Fuzzies from the First Lady."

151 **"When it comes to women"**: Dowd, "Candidate's Wife; Hillary Clinton as Aspiring First Lady."

152 **"unmasked as a counterfeit feminist"**: Maureen Dowd, "Cowboy Feminism," *New York Times*, April 11, 1999, http://www.nytimes.com/1999/04/11/opinion/liberties-cowboy-feminism.html.

152 **Her marriage was a farce:** Sheehy, "What Hillary Wants."

153 **"a shrew whose capacity for denial"**: Tapper, "The Clinton Marriage."

153 **Bill Clinton allegedly rejected a drafted speech:** Dowd, "Monica Gets Her Man."

153 **It turned out that liberal, successful professional women:** Kate Kelly, "Meet the Smart New York Women Who Can't Stand Hillary Clinton," *New York Observer*, January 17, 2000, http://observer.com/2000/01/meet-the-smart-new-york-women-who-cant-stand-hillary-clinton/.

153 **"During my time in Washington, I heard Hillary Clinton called many things"**: Doug Thompson, "Yes, Hillary Clinton Is a Bitch," *Capitol Hill Blue*, November 16, 2007, http://www.capitolhillblue.com/node/3815.

154 **"a very female style of comedy"**: Yael Kohen, interview with the author, March 3, 2017.

154 **"broad and physical role"**: Liza Mundy, "Why Janet Reno Fascinates, Confounds and Even Terrifies America?" *Washington Post*, January 25, 1998, http://www.washingtonpost.com/wp-srv/politics/govt/admin/stories/reno012598.htm.

154 **"I remember thinking that was kind of bullshit"**: Tina Fey, *Bossypants* (New York: Little, Brown and Company, 2011), 135.

155 **In the "Janet Reno's Dance Party" sketches:** Will Ferrell, "Janet Reno Sketches," *Saturday Night Live*, NBC, https://www.nbc.com/saturday-night-live/cast/will-ferrell-15141/impersonation/janet-reno-89921.

156 **"led with her values"**: Jamie Gorelick, interview with the author, October 2015.

156 **She was called the most qualified of the president's cabi-

net officials: Janet Reno, "Swearing-In Ceremony of Janet Reno as United States Attorney General," YouTube video, March 12, 1993, 8:36, posted by clintonlibrary42, April 6, 2012, https://www.youtube.com/watch?v=1H27Ni6bu1s.

156 **"the most urgent issue I faced":** Janet Reno, "Opening Statement before the Crime Subcommittee of the House Judiciary Committee," *Frontline*, PBS, August 1, 1995, http://www.pbs.org/wgbh/pages/frontline/waco/renoopeningst.html.

157 **"Butcher Reno!":** Joe Rosenbloom III, "Waco: More than Simple Blunders?" *Wall Street Journal*, October 17, 1995, http://www.pbs.org/wgbh/pages/frontline/waco/blunders.html.

157 **"like a pileated woodpecker":** Mundy, "Why Janet Reno Fascinates, Confounds and Even Terrifies America?"

158 **"Selections from the Janet Reno Collection":** Michael Crawford, "Selections from the Janet Reno Collection," Condé Nast, https://condenaststore.com/featured/selections-from-the-janet-reno-collection-michael-crawford.html.

158 **"a self-conscious hunch to her shoulders":** Lincoln Caplan, "Janet Reno's Choice," *New York Times*, May 15, 1994, http://www.nytimes.com/1994/05/15/magazine/janet-reno-s-choice.html.

159 **Rather than talk forcefully:** Caplan, "Janet Reno's Choice."

159 **"slow Florida twang":** Caplan, "Janet Reno's Choice."

160 **"more suited to be the Secretary of Health and Human Services":** Caplan, "Janet Reno's Choice."

160 **"not threatened by a successful woman":** Nancy Gibbs, "Truth, Justice and the Reno Way," *Time*, July 12, 1993, http://content.time.com/time/subscriber/article/0,33009,978865-7,00.html.

160 **"because Janet Reno is her father":** Ed Pilkington, "The Joke That Should Have Sunk McCain," *The Guardian*, September 1, 2008, http://www.theguardian.com/lifeandstyle/2008/sep/02/women.johnmccain.

160 **He allegedly apologized:** Pilkington, "The Joke That Should Have Sunk McCain."

160 **Reno's gayness was so often assumed that she addressed it publicly:** Nick Paumgarten, "Aunt Janny," *New Yorker*, Octo-

ber 1, 2007, http://www.newyorker.com/magazine/2007/10/01
/aunt-janny.

162 **if Reno was a "normal woman":** Mundy, "Why Janet Reno Fascinates, Confounds and Even Terrifies America?"

162 **Some reports from their trip:** Elaine Sciolino, "Madeleine Albright's Audition," *New York Times,* September 22, 1996, http://www.nytimes.com/1996/09/22/magazine/madeleine-albright-s-audition.html?pagewanted=all.

164 **"prompting criticism there that she does not take her job seriously enough":** Sciolino, "Madeleine Albright's Audition."

165 **"snarled" to make her point:** Carol Rosenberg, "'Everybody Loves' Tough Albright She's Embraced For Stand After Cuba Downed Planes," *Miami Herald,* March 22, 1996.

165 **"She is like a bulldog":** Michael Dobbs, "With Albright, Clinton Accepts New U.S. Role," *Washington Post,* December 8, 1996, http://www.washingtonpost.com/wp-srv/politics/govt/admin/stories/albright120896.htm.

165 **"There are very few other members of Clinton's Cabinet":** Michael Dobbs and John M. Goshko, "Albright's Personal Odyssey Shaped Foreign Policy Beliefs," *Washington Post,* December 6, 1996.

166 **"taken out the cojones":** Stanley Meisler, "Cuban Pilots Gloated over Shoot-Down," *Chicago Sun-Times,* February 28, 1996.

166 **A former representative from Venezuela accused her:** Hilary Bowker and Richard Roth, "U.S. Secretary of State Madeleine Albright Is an Outspoken Woman," CNN, December 5, 1996.

166 **"too strident":** "Albright: Not One to Pull Punches," CNN, December 6, 1997, http://www.cnn.com/ALLPOLITICS/1997/9612/06/2albright.roth/index.shtml.

166 **called the treatment of Albright:** Warren Bass, "Cold War," *New Republic,* December 13, 1999, https://newrepublic.com/article/79878/cold-war-holbrooke-albright.

167 **foreign colleagues working late:** Peter J. Boyer, "General Clark's Battles," *New Yorker,* November 17, 2003, http://www.newyorker.com/magazine/2003/11/17/general-clarks-battles.

167 **was also the headline:** Walter Isaacson, "Madeleine's War," *Time,* May 9, 1999.

ENDNOTES 345

168 **She also deployed animal brooches:** Lauren Collins, "Big Pin," *New Yorker*, October 5, 2009, https://www.newyorker.com /magazine/2009/10/05/big-pin.

CHAPTER 7: FEMALE ANGER

172 **"I didn't expect such negativity":** Paula Cole, interview with the author, 2015.

173 **"women play electric guitars":** David Browne, "Tragic Kingdom," *Entertainment Weekly*, August 2, 1996, http://ew.com /article/1996/08/02/tragic-kingdom/.

173 **"the princess of post-adolescent feminist angst":** Neal Karlen, "On Top of Pop but Not with One Voice," *New York Times*, June 29, 1997, http://www.nytimes.com/1997/06/29/arts /on-top-of-pop-but-not-with-one-voice.html.

173 **"little-girl type" or "riot grrrl variety":** Karlen, "On Top of Pop but Not with One Voice."

173 **"Flaunting her bare midriff":** Karlen, "On Top of Pop but Not with One Voice."

174 **the "Nancy Reagan of Lilith Fair":** Chris Wilman, "A Fair to Remember," *Entertainment Weekly*, June 19, 1998, http://www .gdrmusic.com/atnatalie/library/nam/980619.htm.

174 **"In the aftermath of Alanis":** "Fiona Apple, 'Tidal'" in "100 Best Albums of the '90s," *Rolling Stone*, April 27, 2011, http:// www.rollingstone.com/music/lists/100-best-albums-of-the -nineties-20110427/fiona-apple-tidal-20110516.

174 **"vocal flutter-kicks":** Karen Schoemer, Yahlin Chang, and Kate Cambor, "Quiet Grrls," *Newsweek*, June 30, 1997.

174 **"twirling banshee . . . roaming the stage wildly":** "The People Column," *Miami Herald*, March 24, 1998.

174 **"grand drama in private dilemmas":** Jon Pareles, "Hold the Anger, Please," *New York Times*, December 27, 1996, http:// www.nytimes.com/1996/12/27/arts/hold-the-anger-please .html.

174 **"unleashed the hellish fury":** Mike Boehm, "Pop Music Review; Paula Cole: Impressive Range, Emotions Ripe for a Rock Diva," *Los Angeles Times*, October 28, 1997.

175 **"little pile of rags":** Wilman, "A Fair to Remember."

176 **grossing $16.5 million:** Timothy Finn, "For McLachlan, All's Fair Second Time Around," *Newark Star-Ledger*, July 7, 1998.

176 **"Call us insensitive":** Schoemer, Chang, and Cambor, "Quiet Grrls."

177 **Activists like Hanna:** *The Punk Singer*, directed by Sini Anderson (Opening Band Films, 2013).

177 **"Not only do we live in a totally fucked-up patriarchal society":** Sara Marcus, *Girls to the Front: The True Story of the Riot Grrrl Revolution*, Kindle edition (New York: HarperCollins, 2010), loc. 3144.

178 **was inspired to join the movement:** "Revolution, Girl Style," *Newsweek*, November 22, 1992, http://www.newsweek.com/revolution-girl-style-196998.

178 **"It felt like finally jumping into a lake":** Sara Marcus, interview with the author, April 2016.

179 **"feminist fury":** Marcus, *Girls to the Front*, loc. 3144.

179 **"militant slant":** Nina Malkin, "It's A Grrrl Thing," *Seventeen*, May 1993.

180 **"Alternative culture in the 90s":** Sarah Seltzer, "'Bitch' Founder Andi Zeisler on a Battle That Remains Only Half-Won," *Flavorwire*, April 19, 2016, http://flavorwire.com/568635/bitch-founder-andi-zeisler-on-a-battle-that-remains-only-half-won.

181 **"a major breakthrough in the expression":** Nataki H. Goodall, "Depend on Myself: T.L.C. and the Evolution of Black Female Rap," *Journal of Negro History* 79, no. 1 (Winter 1994).

182 **gun violence and HIV/AIDS:** Arthur L. Kellermann, Dawna Fuqua-Whitley, and Constance S. Parramore, *Reducing Gun Violence: Community Problem Solving in Atlanta* (Washington, DC: National Institute of Justice, June 2006).

182 **their hometown, Atlanta:** "The First Safest/Most Dangerous City Listing," *Morgan Quitno Press*, 1995, accessed November 3, 2017, http://www.morganquitno.com/1st_safest.htm.

182 **"I like it when you":** Dallas Austin and Lisa "Left Eye" Lopes, "Ain't 2 Proud 2 Beg," AZLyrics, accessed November 3, 2017, http://www.azlyrics.com/lyrics/tlc/aint2proud2beg.html.

182 **"There aren't enough positive ones out there"**: Lenny Stoute, "Fun-Feminist TLC Trio Rappin' on Boyz Door," *Toronto Star*, July 30, 1992.

183 **The only female pop act**: Kenneth Partridge, "TLC's CrazySexyCool at 20: Classic Track-by-Track Album Review," *Billboard*, November 15, 2014, http://www.billboard.com/articles /review/album-review/6319789/tlcs-crazysexycool-at-20-classic -track-by-track-album-review.

183 **a trick to hook listeners**: "People," *Dallas Morning News*, December 31, 1994.

183 **"eliminate the condoms and clean up the lyrics"**: Dennis Hunt, "TLC: Condom Fashions Are a Political Statement," *Los Angeles Times*, April 26, 1992, http://articles.latimes.com/1992 -04-26/entertainment/ca-1225_1_fashion-statement.

183 **"cartoonish-looking trio"**: Sonia Murray, "'CrazySexyCool' TLC toys with a little bit of everything," *Atlanta Journal-Constitution*, November 14, 1994.

183 **"standing up to macho men"**: William Plummer, "In the Heat of the Night," *People*, June 27, 1994, http://people.com/archive /in-the-heat-of-the-night-vol-41-no-24/.

183 **"the girls"**: Bill Diggins, email with the author, June 2015.

184 **"Seems that Left-Eye"**: Curtis Peck, "Morning Briefing," *St. Louis Post-Dispatch*, June 11, 1994, https://www.newspapers .com/newspage/142399388/.

184 **"combustive"**: Plummer, "In the Heat of the Night."

185 **"Strong child-woman"**: Joan Morgan, "The Fire This Time," *Vibe*, November 1994.

186 **"It's so backward"**: Morgan, "The Fire This Time."

186 **"No doubt Rison is hoping"**: "Some Like It Hot; Pop Group TLC Love to Shock Their Fans," *Daily Record*, May 6, 1995.

186 **"TLC burns up the charts"**: *Behind the Music*, season 2, episode 26, "TLC," directed by Nicholas Caprio, aired April 18, 1999, on VH1.

186 **"Andre has some hellraiser in him"**: Len Pasquarelli, "Trouble a Rison Mainstay Off-Field Excesses Shadow Otherwise Brilliant Career," *Atlanta Journal-Constitution*, June 10, 1994.

186 **"After almost three days of shopping"**: Morgan, "The Fire This Time."

186   **"the number one bad girls of pop":** *Behind the Music,* "TLC."

187   **the first girl group to sell:** Partridge, "TLC's CrazySexyCool at 20."

187   **only group member:** Steve Huey, review of *Oooooooohhh . . . On the TLC Tip,* AllMusic, accessed November 3, 2017, http://www.all-music.com/album/oooooooohhhon-the-tlc-tip-mw0000678115.

188   **"a monster":** Diana E. Lundin, "Clubbing Brenda W: Anti-Fans Out to Get Villainess of '90210,'" *Los Angeles Daily News,* February 8, 1993.

188   **"snobbery, hostility and general brattiness":** Mark Ehrman, "Cliques: The Importance of Hating Brenda," *Los Angeles Times,* February 7, 1993.

189   **She was reportedly financially unstable:** Georgea Kovanis, "On- and Off-Screen, '90210' Actress Shannen Doherty Is Taking Her Lumps," *Tulsa World,* March 7, 1993.

189   **"One bad apple":** Kovanis, "On- and Off-Screen, '90210' Actress Shannen Doherty Is Taking Her Lumps."

189   **Brenda hate became a cottage industry:** Kovanis, "On- and Off-Screen, '90210' Actress Shannen Doherty Is Taking Her Lumps."

189   **"wonderfully nasty tattle sheet":** J. D. Considine, "Hating Brenda of '90210' Begets a Small Industry," *Baltimore Sun,* August 8, 1993, http://articles.baltimoresun.com/1993-08-17/features/1993229160_1_brenda-morataya-darby.

190   **"I'm not saying I don't have my moments":** Lundin, "Clubbing Brenda W."

190   **"Why is it when a man":** "Salem Bitch Trial," *Saturday Night Live Transcripts,* accessed November 3, 2017, http://snltranscripts.jt.org/93/93bbitch.phtml.

190   **what a bad influence she was:** J. D. Reed, "A Life on the Edge," *People,* June 14, 1993, http://people.com/archive/cover-story-a-life-on-the-edge-vol-39-no-23/.

190   **reportedly started a fistfight:** Corey Sinclair, "Tori Spelling Admits to Role in Axing of Shannen Doherty from Beverly Hills 90210," *News Corp Australia Network,* October 6, 2015.

191   **"Shock your public":** Lisa Schwarzbaum, "Shannen Doherty:

Image RX," *Entertainment Weekly*, April 23, 1993, http://ew.com
/article/1993/04/23/shannen-doherty-image-rx/.

CHAPTER 8: **MANLY**

193   **A 2015 study of gender and anger:** Arizona State University,
       "Study Shows Angry Men Gain Influence and Angry Women
       Lose Influence," news release, October 27, 2015, https://asunow
       .asu.edu/20151027-study-shows-angry-men-gain-influence
       -and-angry-women-lose-influence.

194   **"tough, outspoken women":** Kenneth R. Clark, "In Their
       Prime: On TV This Fall, It's the Year of the 90s Woman," *Chi-
       cago Tribune*, November 29, 1992.

195   **"unapologetically vacillates between being fat":** O'Connor,
       "By Any Name, Roseanne Is Roseanne Is Roseanne."

195   **"Quasimodo on a bad night":** Ryan Murphy, "Fashion's Mean-
       est Man: The Watcher of the Worst Dressed Talks of Hollywood
       Horror, Nancy Reagan and Guilty Pleasures," *Chicago Tribune*,
       January 9, 1991.

195   **"trying to overfeed":** Chrissy Iley, "Fierce Creature," *The
       Guardian*, October 7, 2008, http://www.theguardian.com
       /lifeandstyle/2008/oct/08/celebrity.comedy.

195   **"make many prominent feminists uneasy":** O'Connor, "By
       Any Name, Roseanne Is Roseanne Is Roseanne."

196   **"Why do you think that everyone thinks":** *Kurt Cobain: Mon-
       tage of Heck*, directed by Brett Morgen (HBO Documentary
       Films, 2015).

196   **"gyrating in G-string":** Daphne Merkin, "Endless Love," in
       The Fame Lunches (New York: Farrar, Straus, and Giroux, 2014).

196   **"a curdled version of the all-American girl":** Laura Barton,
       "Love Me Do," *The Guardian*, December 11, 2006, https://www
       .theguardian.com/film/2006/dec/11/biography.popandrock.

197   **"I remember the jokes backstage":** Amy Finnerty, interview
       with the author, January 14, 2016.

197   **"No wonder Courtney Love behaved":** Angela Buttolph, "Am I
       Too Old to Wear a Baby Doll?" *Evening Standard*, October 19, 2001.

197 **"young, gothy Bette Midler"**: Barbara Ellen, "Interview: Courtney Love: Love and Death and the Hole Damn Thing," *New York Observer,* November 15, 1998.

197 **"taking up public space reserved"**: Kylie Murphy, "'I'm Sorry—I'm Not Really Sorry': Courtney Love and Notions of Authenticity (Focus on Younger Women)," *Hecate* 27, no.1 (May 1, 2001).

197 **"we're not used to seeing in a woman"**: Kevin Sessums, "Love Child," *Vanity Fair,* June 1995, http://www.vanityfair.com/news /1995/06/courtney-love-199506.

198 **"How many drug-addled rock stars"**: Kim France, "The New Courtney," *Slate,* April 13, 1997, http://www.slate.com/articles /news_and_politics/assessment/1997/04/the_new_courtney .html.

198 **"notoriously vengeful"**: "Spin's New Spin on Courtney Cover," *New York Post,* October 18, 1998.

199 **"She felt like I had some beef against her"**: Tabitha Soren, interview with the author, April 2015.

199 **"She lets herself go there"**: Brett Morgen, interview with the author, 2015.

199 **"I can't compare it to rape"**: Kim France, "Feminism Amplified," *New York,* June 3, 1996, https://books.google.com /books?id=1eICAAAAMBAJ&pg=PA34&dq=kim+france +feminism+rocks&hl=en&sa=X&ved=0CCkQ6AEwA2oVCh MIq7Pltv_TxwIVyFg-Ch1axwqA#v=onepage&q&f=false.

200 **readers called her a "vile hag"**: *Kurt Cobain: Montage of Heck,* directed by Morgen.

200 **"his talent and his wounded sweetness"**: Michael O'Sullivan, "Documentary: Love Hurts," *Washington Post,* July 17, 1998, http://www.washingtonpost.com/wp-srv/style/longterm /movies/videos/kurtandcourtneyosullivan.htm.

200 **"three inches taller than he was"**: Charles R. Cross, "The Moment Kurt Cobain Met Courtney Love," *Daily Beast,* April 5, 2014, http://www.thedailybeast.com/articles/2014/03/04/the -moment-kurt-cobain-met-courtney-love.html.

200 **She still contends with**: Heather Saul, "Courtney Love's Moving Tribute to Kurt Cobain on Instagram Hit by Abu-

sive Comments," *The Independent*, August 25, 2015, http://
www.independent.co.uk/news/people/courtney-loves-moving
-tribute-to-kurt-cobain-on-instagram-hit-by-abusive-com
ments-10470453.html.

201 **"denigration of the anthem"**: Barry M. Horstman, "The Boos
Swell: Roseanne in Rockets' Red Glare," *Los Angeles Times*,
July 27, 1990.

201 **Condemnations included:** Russell Baker, "We Need a Star-
Spangled Singing Test," *Sarasota Herald-Tribune*, August 4, 1990,
https://news.google.com/newspapers?nid=1755&dat=1990
0804&id=CLseAAAAIBAJ&sjid=vXoEAAAAIBAJ&pg=32
56,4271867&hl=en.

201 **Opera star Robert Merrill:** Horstman, "The Boos Swell."

202 **"bossy, and less effective than male counterparts"**: Jillian
Kramer, "Female Leaders Who Behave Like Men Are Seen as
Bossy," Glamour.com, December 6, 2016, https://www.glamour
.com/story/female-leaders-who-behave-like-men-are-seen-as
-bossy.

202 **When the show hit number one:** Roseanne Barr, "And I
Should Know," *New York*, May 2011, http://nymag.com/arts/tv
/upfronts/2011/roseanne-barr-2011-5/index2.html.

203 **"Rage is Roseanne's ozone":** Iley, "Fierce Creature."

203 **detractors dubbed her:** Plotz, "Domestic Goddess Dethroned."

203 **"Women still take her more seriously":** Eric Weinrib, conversation with the author at Tribeca Film Festival 2015.

204 **"Did you actually get to say something nice":** "Nirvana-Short
Interview + MTV 1994 Year in Rock Report-12/23/94," MTV,
YouTube video, 7:00, posted by NirvanaForevermore, April 5,
2013, https://www.youtube.com/watch?v=fAX23jf2seU.

204 **"Courtney was interested in the canon":** Kim France, interview with the author, September 2, 2015.

204 **"contain her huge appetites":** Merkin, "Endless Love."

204 **"too mammoth to be confined to a genre":** Murphy, "'I'm
Sorry—I'm Not Really Sorry.'"

204 **"the kind of ambition":** France, "Feminism Amplified."

204 **"Love's vilification as a bitch":** Murphy, "'I'm Sorry—I'm Not
Really Sorry.'"

205   **"She knew far more"**: Cross, "The Moment Kurt Cobain Met Courtney Love."

206   **"It seems a pity that Love"**: Merkin, "Endless Love."

206   **"whitewash her tarnished image"**: O'Sullivan, "Documentary: Love Hurts."

207   **"I thought, like Madonna"**: Nancy Jo Sales, "Love in a Cold Climate," *Vanity Fair,* November 2011, http://www.vanityfair.com /news/2011/11/courtney-love-201111.

207   **"When you say 'rock star'"**: Adam Diehl, interview with the author, May 2016.

CHAPTER 9: DAMAGED GOODS

209   **in one 2005 study, 15 percent**: A. Laye-Gindhu and K. A. Schonert-Reichl, "Nonsuicidal Self-Harm Among Community Adolescents: Understanding the 'Whats' and 'Whys' of Self-Harm," *Journal of Youth and Adolescents* 34, no. 5 (2005): 447–57, doi:10.1007/s0964-005-7262-z.

209   **Boys were more likely to**: Patrick L. Kerr, Jennifer J. Muehlenkamp, and James M. Turner, "Nonsuicidal Self-Injury: A Review of Current Research for Family Medicine and Primary Care Physicians," *Journal of the American Board of Family Medicine* 23, no. 2 (March–April 2010): 240–259, doi: 10.3122 /jabfm.2010.02.090110.

209   **never treated a girl who cut herself**: Mary Pipher, *Reviving Ophelia: Saving the Selves of Adolescent Girls* (New York: Riverhead Books, 1994), 157.

210   **"Self-mutilation may well be a reaction"**: Pipher, *Reviving Ophelia*, 157.

210   **nearly 17 percent of girls**: David M. Cutler, Edward L. Glaeser, and Karen E. Norberg, "Explaining the Rise in Youth Suicide," in *Risky Behavior among Youths: An Economic Analysis*, ed. Jonathan Gruber (University of Chicago Press, 2001), 219–270, http://www.nber.org/chapters/c10690.pdf.

211   **wrote about cutting herself**: Atoosa Rubenstein, "Sad Dad Could Be Depressed," *Desert News*, August 14, 2006.

211   **"the new bulimia"**: Linda Holler, *Erotic Morality: The Role of*

Touch in Moral (New Brunswick, NJ: Rutgers University Press, 2002).

211 **she admitted to cutting:** Michelle Green, "True Confessions," *People*, December 4, 1995, http://people.com/archive/cover-story -true-confessions-vol-44-no-23/.

212 **"Self-injury can be understood":** Natasha Alexander and Linda Clare, "You Still Feel Different: The Experience and Meaning of Women's Self-Injury in the Context of a Lesbian or Bisexual Identity," *Journal of Community & Applied Social Psychology* 14, no. 2 (March–April 2004): 70–84.

213 **"heroin-chic":** Christine L. Harold, "Tracking Heroin Chic: The Abject Body Reconfigures the Rational Argument," *Argumentation and Advocacy* 36, no. 2 (Fall 1999): 65–76, http://eric .ed.gov/?id=EJ592792.

213 **"like an underfed Calvin Klein model":** Sasha Frere-Jones, "Extraordinary Measures," *New Yorker*, October 10, 2005, http:// www.newyorker.com/magazine/2005/10/10/extraordinary -measures.

213 **"Kate Moss with songs":** Richard Harrington, "Fiona Apple: The Time Is Ripe," *Washington Post*, November 28, 1999, http:// www.washingtonpost.com/wp-srv/WPcap/1999-11/28/003r -112899-idx.html.

213 **"teenager's sense of drama":** Frere-Jones, "Extraordinary Measures."

213 **"Apple has often seemed":** Dave Tianan, "Review: Apple sings from a core of emotion," *Milwaukee Journal Sentinel*, August 13, 2006.

213 *Time* **named the video:** Katy Steinmetz, "Top 10 Controversial Music Videos: 'Criminal,'" *Time*, June 6, 2011, http://entertain ment.time.com/2011/06/07/top-10-controversial-music-videos/.

213 **"overtones of child porn":** Ben Williams, "A World-Class Drama Queen," *New York*, October 10, 2005, http://nymag.com /nymetro/arts/music/pop/14612/.

214 **"a self-obsessed drama queen":** Harrington, "Fiona Apple: The Time Is Ripe."

214 **"a tsunami of adolescent feelings":** Harrington, "Fiona Apple: The Time is Ripe."

214 **"might lead one to believe"**: Chris Heath, "Fiona: The Caged Bird Sings," *Rolling Stone*, January 22, 1998, http://www.rolling stone.com/music/news/fiona-the-caged-bird-sings-19980122.

214 **"For me, it wasn't about getting thin"**: Heath, "Fiona: The Caged Bird Sings."

216 **"They kept saying she hurt her thigh"**: Kathy O'Hearn, interview with the author, October 2015.

217 **seven hundred fifty million people**: "More Information About: Prince Charles and Lady Diana Spencer's Wedding," BBC, accessed November 6, 2017, http://www.bbc.co.uk/history/events /prince_charles_and_lady_diana_spencers_wedding.

217 **"modified bowl haircut"**: Warren Hoge, "Diana, Princess of Wales, 36, Dies in a Crash in Paris," *New York Times*, August 31, 1997, http://www.nytimes.com/1997/08/31/world/europe/diana -obit.html.

217 **"always pitched out front"**: "Princess Diana Interview Part 2," *Panorama*, BBC, YouTube video, aired November 1995, 9:54, posted by CAMELOTHSPENCER, October 22, 2008, https:// www.youtube.com/watch?v=Zlxs_JG1dDA.

218 **"adored fancy clothes"**: Sarah Lyall, "Charles and Diana Agree on Divorce Terms," *New York Times*, July 13, 1996, http://www .nytimes.com/1996/07/13/world/charles-and-diana-agree-on -divorce-terms.html?pagewanted=all.

218 **"thick as a plank"**: "Princess Di Admits to Being 'Thick as a Plank,'" *Ocala Star-Banner*, January 21, 1987, https://news .google.com/newspapers?nid=1356&dat=19870121&id= V8BPAAAAIBAJ&sjid=sgYEAAAAIBAJ&pg=5271,2024739 &hl=en.

218 **He charged that the royal family**: Andrew Morton, *Diana: Her True Story* (New York: Simon & Schuster, 1992), 85.

219 **"the cause of the marriage problems"**: Morton, *Diana*, 73.

219 **bulimia rates tripled**: Laura Currin, Ulrike Schmidt, Janet Treasure, and Herschel Jick, "Time Trends in Eating Disorder Incidence," *British Journal of Psychiatry* 186, no. 2 (January 2005): 132–135, http://bjp.rcpsych.org/content/186/2/132.

219 **Charles and Diana were immersed**: Hoge, "Diana, Princess of Wales, 36, Dies in a Crash in Paris."

220 **Diana "posed knowingly on Mediterranean holidays":** Hoge, "Diana, Princess of Wales, 36, Dies in a Crash in Paris."

220 **"I never know where a lens is going to be":** "Princess Diana Interview Part 2," *Panorama.*

220 **"paranoid and foolish":** Morton, *Diana*, 21.

221 **"Vengeance is a dish best served cold":** Nicholas Wapshott, "Diana: Hunted or Huntress?" *Newsweek*, October 31, 2013, http://www.newsweek.com/2013/11/01/diana-hunted-or-hunt ress-243882.html.

221 **men, who comprised more than half of journalists:** Anna Griffin, "Where Are the Women?" *Nieman Reports*, September 11, 2015.

CHAPTER 10: VICTIMS AND VIOLENCE

223 **"cajole, demand, infiltrate":** Harry F. Waters, "Whip Me, Beat Me . . . and give me great ratings. A network obsession with women in danger," *Newsweek*, November 11, 1991.

224 **"faces the loss of her practice":** Waters, "Whip Me, Beat Me."

224 **250 made-for-TV movies in the 1992 season:** Waters, "Whip Me, Beat Me."

224 **"Somewhere in America":** Mark Harris, "Mad Women on TV," *Entertainment Weekly*, April 24, 1992, http://ew.com/article /1992/04/24/mad-women-tv/.

225 **"the first thing many say is":** Waters, "Whip Me, Beat Me."

225 **"Women are being beaten":** Mike Duffy, "Women, TV and Ultraviolence: Women Are the Victims, Stars and Consumers of Increasingly Gory Shows," *Detroit Free Press*, October 4, 2005.

225 **"It does seem like there's a shocking amount":** Molly Willow, "Vivid Violence: More TV Shows Using Grisly Crimes against Women to Draw Audiences," *Columbus Dispatch*, November 8, 2005.

225 **"When I have these meetings":** Deborah Hastings, "Lifetime's 'Attitudes' Tackles Feminism," Associated Press, April 21, 1992.

226 **"I don't get it. It's two bitches in a car":** Becky Aikman, *Off the Cliff: How the Making of Thelma & Louise Drove Hollywood to the Edge* (New York: Penguin Publishing Group, 2017), 3–4.

226 **"leads to increased acceptance of rape"**: Mike Feinsilber, "Psychologists: TV Ignores, Distorts Lives of Most People," Associated Press, February 25, 1992.

226 **"We look to the audience"**: Willow, "Vivid Violence."

227 **"teen girl psychopath"**: Diana Jean Schemo, "Hidden and Haunted behind the Headlines," *New York Times*, June 12, 1992, http://www.nytimes.com/1992/06/12/nyregion/hidden -haunted-behind-headlines-parents-accused-long-island-teen -ager-are.html?pagewanted=all.

227 **"Fatal Attraction . . . teenage style"**: Joe Treen, "Treachery in the Suburbs," *People*, June 29, 1992, http://people.com/archive /treachery-in-the-suburbs-vol-37-no-25/.

227 **for $8,000**: Richard Cohen, "The Name of De Rosa," *Washington Post*, August 30, 1992, https://www.washingtonpost .com/archive/lifestyle/magazine/1992/08/30/the-name-of-de -rosa/0f846f1c-3614-409f-8dab-225943c213bd/?utm_term =.a8e5ad3a5a99.

227 **"wore cutoff jeans"**: "'The Long Island Lolita,'" *Newsweek*, June 14, 1992, http://www.newsweek.com/long-island-lolita -199280.

227 **"vilified in print as a venal, spoiled little bitch"**: Steve Dunleavy, "Lolita's Dad Begs Forgiveness for His Girl," *New York Post*, February 3, 1999, http://nypost.com/1999/02/03/lolitas-dad -begs-forgiveness-for-his-girl-amys-dad-pleads-for-forgiveness -exclusive/.

228 **"shrewd, manipulative, and brazen"**: Diana Jean Schemo, "Not-Guilty Plea Entered by Teen-Ager in Shooting," *New York Times*, June 3, 1992, http://www.nytimes.com/1992/06/03 /nyregion/not-guilty-plea-entered-by-teen-ager-in-shooting .html.

228 **"old enough to be her father"**: William A. Henry, "Read All About Lolita!" *Time*, June 15, 1992, http://content.time.com /time/magazine/article/0,9171,975772,00.html.

228 **"To call her a seventeen-year-old girl"**: Schemo, "Not-Guilty Plea Entered by Teen-Ager in Shooting."

228 **nearly seven years**: John T. McQuiston, "Amy Fisher Is Released After Almost 7 Years in Prison," *New York Times*, May 11, 1999,

http://www.nytimes.com/1999/05/11/nyregion/amy-fisher
-is-released-after-almost-7-years-in-prison.html.

228 **severe depression and attempted suicide twice:** Joe
Treen, "Sex, Lies, and Videotapes," *People*, October 12, 1992,
http://people.com/archive/cover-story-sex-lies-videotapes
-vol-38-no-15/.

228 **"So here I was, on the brink of sweet sixteen":** "Amy Fisher
Tells of Sex Abuse," *Orlando Sentinel*, April 4, 1993, http://
articles.orlandosentinel.com/1993-04-04/news/9304
040004_1_amy-fisher-joey-buttafuoco-abortion.

229 **"He's going to say whatever he needs to":** Lesléa Newman,
interview with the author, July 2015.

229 **"If you become a teenage prostitute":** John O'Connor, "Critic's
Notebook; The Line Between Dramas And Lies," *New York Times*,
December 31, 1992, http://www.nytimes.com/1992/12/31/arts
/critic-s-notebook-the-line-between-dramas-and-lies.html.

229 **"a $180-a-night prostitute":** Treen, "Treachery in the Sub-
urbs."

229 **Amy Fisher Bang for Your Bucks:** "In Living Color-Jim
Carrey does Joey Buttafuoco, Amy Fisher bang for you bucks
seminar," *In Living Color*, Fox, YouTube video, 2:47, posted
by vinnyqua, December 12, 2009, https://www.youtube.com
/watch?v=JW6NbI6dSx0.

230 **roughly triple the amount Fisher did:** Diane Ketcham,
"About Long Island; 3 TV Films, 3 Versions of Amy Fisher
Case," *New York Times*, December 6, 1992, http://www.nytimes
.com/1992/12/06/nyregion/about-long-island-3-tv-films-3
-versions-of-amy-fisher-case.html?pagewanted=all.

230 **"chastened hussy":** Dan Barry, "The Nation: No Way Out; Still
Gawking After All These Years," *New York Times*, May 16, 1999,
http://www.nytimes.com/1999/05/16/weekinreview/the
-nation-no-way-out-still-gawking-after-all-these-years.html.

230 **"classy tear-away underwear":** "Saturday Night Live: Amy
Fisher," *Saturday Night Live*, NBC, Hulu video, 2:07, http://www
.hulu.com/watch/270886.

230 **A 2008 report compiled by the Department of Justice:** Margaret A. Zahn et al., "Violence by Teenage Girls: Trends and

Context," US Department of Justice, May 2008, https://www
.ncjrs.gov/pdffiles1/ojjdp/218905.pdf.

230 **girls were more likely to perpetrate:** T. M. Franke, A. L. T.
Huynh-Hohnbaum, and Y. Chung, "Adolescent Violence: With
Whom They Fight and Where," *Journal of Ethnic and Cultural
Diversity in Social Work* 11, no. 3–4 (2002).

231 **A 1997 study:** S. Artz, "On Becoming an Object," *Journal of
Child and Youth Care* 11, no. 2 (1997): 17–37.

232 **gave away cocktail wienies and Slice soda:** Melissa Jeltson,
"Lorena Bobbitt Is Done Being Your Punchline," *Huffington
Post*, December 22, 2016, http://www.huffingtonpost.com/entry
/lorena-bobbitt-domestic-violence_us_585ab844e4b0eb58648
4cea9.

232 **"cut heard round the world":** "The Cut Heard Round The
World," *Newsweek*, October 17, 1993, http://www.newsweek.com
/cut-heard-round-world-194218.

232 **"a wake-up call":** Robin Abcarian, "Let's Not Make Lorena
Bobbitt a Feminist Poster Child," *Los Angeles Times*, December 5, 1993, http://articles.latimes.com/1993-12-05/news/vw
-64250_1_lorena-bobbitt.

232 **County police had answered domestic violence calls:**
Lorena Bobbitt, "Lorena Bobbitt '93 Exclusive: Part 1,"
ABC News video, December 5, 1993, accessed September
2016, http://abcnews.go.com/2020/video/lorena-bobbitt-93
-exclusive-scared-11747248.

233 **her 2003 report:** Elizabeth K. Carll, "Violence and Women: News
Coverage of Victims and Perpetrators," *American Behavioral Scientist*, August 1, 2003, http://docplayer.net/24734748-Violence
-and-women-news-coverage-of-victims-and-perpetrators
.html.

233 **"You do not have the right to kill or maim someone":** "No
Tears for Lorena," *Newsweek*, January 23, 1994, http://www
.newsweek.com/no-tears-lorena-187404.

233 **"a symbol of female rage":** Abcarian, "Let's Not Make Lorena
Bobbitt a Feminist Poster Child."

234 **"If . . . she had killed him instead":** Lorena Bobbitt, "Lorena
Bobbitt '93 Exclusive: Part 2," ABC News video, December 5,

1993, http://abcnews.go.com/2020/video/lorena-bobbitt-93
-exclusive-part-11747291.

234 **"Her abuse of him was so barbaric"**: Abcarian, "Let's Not
Make Lorena Bobbitt a Feminist Poster Child."

234 *National Lampoon's* **twenty-fifth anniversary**: Jonathan
Taylor, "He Never Gave Me Orgasm: The Lenora Babbitt
Story," *Variety*, August 19, 1994, http://variety.com/1994/film
/reviews/he-never-give-me-orgasm-the-lenora-babbitt-story
-1200438184/.

235 **"vindictive and said"**: Michael Ross, "Lorena Bobbitt's Trial
for Cutting Penis Begins," *Los Angeles Times*, January 11, 1994,
http://articles.latimes.com/1994-01-11/news/mn-10704_1
_lorena-bobbitt.

235 **"instant feminist pin-up girl"**: Mona Charen, "Lorena Bobbitt, America's Feminist Pin-Up," *Baltimore Sun*, November 15,
1993, http://articles.baltimoresun.com/1993-11-15/news/1993
319151_1_lorena-bobbitt-john-bobbitt-spinning-the-story.

235 **"every woman's fantasy"**: Charen, "Lorena Bobbitt, America's
Feminist Pin-Up."

236 **"The female criminal violates two laws"**: Katherine Dunn,
"Just as Fierce," *Mother Jones*, November/December 1994.

236 **"zipped around Brentwood"**: "Nicole Brown Simpson,"
Biography.com, last updated February 11, 2016, http://www
.biography.com/people/nicole-brown-simpson-21254807.

237 **picked up checks**: Sheila Weller, *Raging Heart: The Intimate
Story of the Tragic Marriage of O. J. and Nicole Brown Simpson*
(Los Angeles: Graymalkin Media, 2016).

237 **"Nicole wanted her space"**: Weller, *Raging Heart.*

237 **"He's going to beat the shit out of me"**: Jeffrey Toobin, *The
Run of His Life: The People v. O. J. Simpson* (New York: Random
House, 1996), 131.

237 **"a terrible joke"**: Sara Rimer, "The Simpson Case: The Marriage; Handling of 1989 Wife-Beating Case Was a 'Terrible
Joke,' Prosecutor Says," *New York Times*, June 18, 1994, http://
www.nytimes.com/1994/06/18/us/simpson-case-marriage
-handling-1989-wife-beating-case-was-terrible-joke.html.

238 **"We regard it as a private matter"**: Roger Simon, "Simp-

son Lost Hero Status in 1989, Not Last Week," *Baltimore Sun*, June 24, 1994, http://articles.baltimoresun.com/1994-06-24 /news/1994175031_1_simpson-american-hero-wife.

238 **"It was really a bum rap":** Josh Meyer, "Police Records Detail 1989 Beating That Led to Charge: Violence: A Bloodied Nicole Simpson, Hiding in Bushes After 911 Call, Told Officers: 'He's Going to Kill Me.' Judge Overruled Prosecutors' Request That Simpson Serve Jail Time," *Los Angeles Times*, June 17, 1994, http:// articles.latimes.com/1994-06-17/news/mn-5290_1_jail-time.

238 **A sushi bar manager recalled:** Weller, *Raging Heart.*

239 **Even friends claimed that Brown:** Sheila Weller, "How O. J. and Nicole Brown's Friends Coped with Murder in Their Midst," *Vanity Fair*, June 12, 2014.

239 **"a crime of passion":** Weller, "How O. J. and Nicole Brown's Friends Coped with Murder in Their Midst."

239 **1998 report on intimate violence:** Lawrence Greenfeld et al., *Violence by Intimates: Analysis of Data on Crimes by Current or Former Spouses, Boyfriends, and Girlfriends* (Washington, DC: Bureau of Justice Statistics, March 1998), https://bjs.gov /content/pub/pdf/vi.pdf.

240 **"a wife beater turned killer":** Ellen Willis, "The Wrath of Clark," *New York Times*, June 15, 1997, http://www.nytimes.com /1997/06/15/books/the-wrath-of-clark.html?pagewanted=all.

240 **"emotional resistance":** Willis, "The Wrath of Clark."

240 **"very, very risky":** "Prosecution Begins Opening Statements in Simpson Trial," ABC News, January 24, 1995.

241 **governors in Maryland and Ohio:** "No Tears for Lorena," *Newsweek.*

241 **"misleading and potentially harmful":** Mary Ann Dutton, Sue Osthoff, and Melissa Dichter, "Update of the 'Battered Women Syndrome' Critique," VAWNet: The National Online Resource Center on Violence against Women, August 2009, http://vawnet .org/material/update-battered-woman-syndrome-critique.

242 **"the case regarding domestic discord":** Jim Hill, "Prosecutors Withdraw New Witnesses in Simpson Trial," CNN, June 20, 1995.

242 **"That's faulty logic":** *Crossfire*, CNN, February 5, 1996.

242 **70 to 80 percent of partner homicides:** Jacquelyn C. Campbell et al., "Assessing Risk Factors for Intimate Partner Homicide," *National Institute of Justice Journal* 250 (2003): 14–19, https://www.ncjrs.gov/pdffiles1/jr000250e.pdf.

243 **"subordinate status in society":** Lori Heise, Mary Ellsberg, and Megan Gottemoeller, "Ending Violence against Women," *Population Reports* (Johns Hopkins University School of Public Health) L, no. 11 (December 1999), https://www.k4health.org/sites/default/files/L%2011.pdf.

243 **ending gender violence was framed:** Heise, Ellsberg, and Gottemoeller, "Ending Violence Against Women."

244 **intimate partners committed fewer murders:** Greenfeld et al., *Violence by Intimates.*

245 **"often literally cowering in fear and shame":** *Violence Against Women: The Communications Evolution* (New York: Avon Foundation for Women, 2012), 4, https://www.multivu.com/assets/46212/documents/Communications-Evolution-Report-original.pdf.

245 **$1 billion to combat domestic abuse:** Laura Meckler, "Five Years after Simpson, War against Domestic Abuse Improves," Associated Press, June 12, 1999, http://onlineathens.com/stories/061399/new_0613990015.shtml#.Waw3uiMrIfF.

245 **$4 billion of government funds:** Lynn Rosenthal, "Sixteen Years of the Violence Against Women Act," White House of President Barack Obama blog, September 23, 2010, https://obamawhitehouse.archives.gov/blog/2010/09/23/sixteen-years-violence-against-women-act.

245 **nearly ten times more likely:** Violence Policy Center, *When Men Murder Women: An Analysis of 2013 Homicide Data* (Washington, DC: September 2015), http://www.vpc.org/studies/wmmw2015.pdf.

245 **increases fivefold:** Dave Gilson, "10 Pro-Gun Myths, Shot Down," *Mother Jones*, January 31, 2013, http://www.motherjones.com/politics/2013/01/pro-gun-myths-fact-check.

245 **it wasn't until 2014:** Travis Waldron, "The NFL's Domestic Violence Policy Isn't Working Because It Wasn't Designed To," *Huffington Post*, October 21, 2016, https://www

.huffingtonpost.com/entry/nfl-josh-brown-domestic-violence
-giants_us_580a1b0be4b02444efa2c5ca.

CHAPTER 11: CATFIGHT

248 **banned from figure skating for life:** Johnette Howard, "Harding Admits Guilt in Plea Bargain, Avoids Prison," *Washington
Post*, March 17, 1994, http://www.washingtonpost.com/wp-srv
/sports/longterm/olympics1998/history/timeline/articles
/time_031794.htm.

248 **people described Nancy:** *The Price of Gold*, directed by Nanette
Burstein (ESPN Films, 2014).

249 **"there was this overriding question":** Nanette Burstein, email
interview with the author, March 6, 2017.

249 **feeding the fantasy:** Abigail Feder, "A Radiant Smile from a
Lovely Lady: Overdetermined Femininity in 'Ladies' Figure
Skating," in *Women on Ice: Feminist Essays on the Tonya Harding/
Nancy Kerrigan Spectacle*, ed. Cynthia Baughman (New York:
Routledge, 1995), 38.

249 **"music box figurine come to life":** Steve Hummer, "Attack on Ice
Princess Marks Disturbing Trend," *Atlanta Journal-Constitution*,
January 8, 1994.

250 **"ice sculpture":** Laura Jacobs, "Pure Desire," in *Women on Ice:
Feminist Essays on the Tonya Harding/Nancy Kerrigan Spectacle*,
ed. Cynthia Baughman (New York: Routledge, 1995).

250 **the best female figure skater in the country:** Jere Longman,
"Jealousy on Ice," *New York Times*, January 6, 1994, http://www
.nytimes.com/packages/html/sports/year_in_sports/01.06
.html.

250 **"a very good patina to her":** *The Price of Gold*, directed by
Burstein.

250 **"erect carriage":** Ellyn Kestnbaum, "What Tonya Harding
Means to Me, or Images of Independent Female Power on Ice,"
in *Women on Ice: Feminist Essays on the Tonya Harding/Nancy Kerrigan Spectacle*, ed. Cynthia Baughman (New York: Routledge,
1995).

250 **"supermodel beautiful":** Larry Mendte, "What Nancy Kerrigan

and Tonya Harding Were Really Like in Lillehammer," *Philadelphia*, February 24, 2014, http://www.phillymag.com/news/2014/02/24/nancy-kerrigan-tonya-harding-attack-lillehammer/.

251 **recalls physical, verbal, and sexual abuse:** Lynda D. Prouse, *The Tonya Tapes* (New York: World Audience, 2008), 57.

251 **her coach paid a competitor five dollars:** *Sharp Edges*, directed by Sandra Luckow (New York: Ojeda Films, 1986).

251 **Her "muscular arms and chunky thighs":** Randall Sullivan, "What's More Fun Than a Good Old Fashioned Tonya Harding Story?" *Rolling Stone*, July 14, 1994.

251 **"Sucking on an asthma inhaler":** Sullivan, "What's More Fun Than A Good Old Fashioned Tonya Harding Story?"

252 **"She didn't play by the rules":** Nanette Burstein, interview with the author, July 23, 2015.

252 **"One of the judges came up to me":** *The Price of Gold*, directed by Burstein.

252 **"illusion of decorous femininity":** Abby Haight, J. E. Vader, and the Staff of the *Oregonian*, *Fire on Ice: The Exclusive Inside Story of Tonya Harding* (New York: Three Rivers Press, 1994), 38.

252 **"looked like she stepped out of a Ralph Lauren catalogue":** Phil Hersh, "Harding, Longtime Coach Different as a Team Can Possibly Be," *Chicago Tribune*, January 25, 1994, http://articles.chicagotribune.com/1994-01-25/sports/9401250188_1_diane-rawlinson-tonya-harding-skate.

253 **"Skating for Tonya is her ticket":** *Sharp Edges*, directed by Luckow.

253 **plenty to jeer at:** Haight, Vader, and the Staff of the *Oregonian*, *Fire on Ice*, 34.

253 **"There was no question":** *The Price of Gold*, directed by Burstein.

253 **perfect technical merit score:** Haight, Vader, and the Staff of the *Oregonian*, *Fire on Ice*.

254 **"If Harding skates a clean program":** Haight, Vader, and the Staff of the *Oregonian*, *Fire on Ice*, 44.

254 **"Her incompetence as a woman":** Feder, "A Radiant Smile from a Lovely Lady."

254 **"brought impurity to the sport":** Jacobs, "Pure Desire."

254 **"Without those jumps she wasn't much to look at"**: Haight, Vader, and the Staff of the *Oregonian*, *Fire on Ice*, 32.

254 **"it reduced her value"**: Feder, "A Radiant Smile from a Lovely Lady," 31.

254 **"rough edges"**: Jere Longman, "Figure Skating; Lines Blur for Kerrigan and Harding," *New York Times*, January 6, 1995, http://www.nytimes.com/1995/01/06/sports/figure-skating -lines-blur-for-kerrigan-and-harding.html.

255 **"powerful welder's arms"**: Mark Starr, "Nancy Kerrigan: 'I'm So Scared,'" *Newsweek*, January 17, 1994, http://www.newsweek .com/nancy-kerrigan-im-so-scared-187512.

255 **"what a crybaby she is"**: "Letters from the People," *St. Louis Post-Dispatch*, January 22, 1994.

256 **"distorted as she watched a life's work"**: Starr, "Nancy Kerrigan: 'I'm So Scared.'"

256 **"We don't want to look at Kerrigan"**: Mark Kiszla, "Kerrigan's Real World Cold as Ice," *Denver Post*, January 9, 1994.

256 **"took a crowbar"**: Hummer, "Attack on Ice Princess Marks a Disturbing Trend."

256 **"defaced a beautiful symbol"**: Kiszla, "Kerrigan's Real World Cold as Ice."

256 **"You want to believe that the beauty"**: Hummer, "Attack on Ice Princess Marks a Disturbing Trend."

256 **"The Olympic princess"**: Kiszla, "Kerrigan's Real World Cold as Ice."

256 **"mistakenly believing she was immune to hate"**: Kiszla, "Kerrigan's Real World Cold as Ice."

256 **"Gillooly said that the hit man"**: Don Fulson, "The Tonya Harding Anniversary Quiz," *Washington Post*, January 29, 1995, https://www.washingtonpost.com/archive/opinions/1995 /01/29/the-tonya-harding-anniversary-quiz/9655a93b-b103 -446f-8586-233136c90111/?utm_term=.db185c9883d7.

256 **"All skating whores will die"**: Sullivan, "What's More Fun Than a Good Old Fashioned Tonya Harding Story?"

257 **"almost-anonymous practitioner"**: Lowell Cohn, "The Sad Truth Behind Attack on Kerrigan," *San Francisco Chronicle*, January 11, 1994.

257 **She became linked:** Julio Laboy, "Tonya's Guard Under Arrest; Cops: Tonya Not Involved in Rink Attack," *Newsday*, January 14, 1994.

258 **Harding and Kerrigan were more intensely covered:** Laurie A. Sheflin, "Ted Koppel Receives IOP Award," *Harvard Crimson*, March 11, 1994, http://www.thecrimson.com/article/1994/3/11/ted-koppel-receives-iop-award-pnetwork/.

258 **Television crews camped out:** Nancy Kerrigan, *Nancy Kerrigan: In My Own Words* (New York: Hyperion, 1996), 55.

258 **a middle-aged resident:** Kevin O'Leary, "Nancys Share the Burden of Fame," *Boston Globe*, February 21, 1994.

259 **"I had to leave home early":** Lynn Harris, interview with the author, June 2015.

260 **"I hope that it's true":** Elizabeth Searle, interview with the author, July 2015.

260 **"never warmed to the combative Harding":** John Jeansonne, "Up Close and Personal," *Newsday*, January 13, 2017.

260 **"a little barracuda":** Johnette Howard, "An Image with Sharp Edges; Skater in Greatest Controversy," *Washington Post*, January 14, 1994.

260 **imagining skating officials' views:** Bob Verdi, "Harding Affair Enough To Spoil Olympian Appetites," *Chicago Tribune*, February 5, 1994.

261 **"Despite my mistakes and my rough edges":** Liz Willen, "Tonya Knew and Didn't Tell; Says She Learned of Cohorts' Roles Soon after Attack on Rival," *Newsday*, January 28, 1994.

261 **"I can't believe she's actually coming":** Christine Brennan, "Logistics, Blame-Laying Concern Skating Now; Separate Kerrigan, Harding Practices Sought," *Washington Post*, February 14, 1994.

261 **"I certainly had not gone to Detroit to win":** Kerrigan, *Nancy Kerrigan: In My Own Words*, 50.

261 **"I worked my butt off":** Christine Brennan and Jim McGee, "Skater Attack Seen as Plot; Competitor's Husband Said to Be a Suspect," *Washington Post*, January 13, 1994.

262 **"dark-hearted sporting she-devil":** Peter Nichols and Dennis Campbell, "Skating's Cold War Has Won Worldwide Media Attention," *Scotland on Sunday*, February 20, 1994.

262 **"It was so rich in its blacks"**: *The Price of Gold*, directed by Burstein.

262 **"peaceful determination"**: Phil Hersh, "Coach Says Kerrigan Way Ahead of Schedule," *Chicago Tribune*, January 20, 1994.

262 **"Nancy was the ice princess"**: Hilary Bauer, interview with the author, July 2015.

262 **"Honestly, I gravitated toward Nancy"**: Jenna Leigh Green, interview with the author, July 2015.

263 **"a media scrum worthier of"**: Nichols and Campbell, "Skating's Cold War Has Won Worldwide Media Attention.

263 **"This was like watching *Dynasty*"**: *The Price of Gold*, directed by Burstein.

264 **"broad streak of bitchiness"**: Sullivan, "What's More Fun Than A Good Old Fashioned Tonya Harding Story?"

264 **"Oh, come on"**: Kim Masters, "Kerrigan Off the Ice Doesn't Seem Half as Nice," *Washington Post*, March 4, 1994, http://www .washingtonpost.com/wp-srv/sports/longterm/olympics1998 /history/timeline/articles/time_030494.htm.

264 giving **"curt answers at press conferences"**: Leigh Montville, "On with the Show," *Sports Illustrated*, December 5, 1994, http:// www.si.com/vault/1994/12/05/132804/on-with-the-show-the -hardest-act-for-nancy-kerrigan-to-follow-on-the-ice-or-off -turns-out-to-be-her-own.

264 **"They're bending lots of rules"**: Phil Hersh, "Figure Skating Surprising, but Champion Isn't," *Chicago Tribune*, February 27, 1994, http://articles.chicagotribune.com/1994-02-27 /sports/9402270374_1_figure-skating-association-jeff-gillooly -nancy-kerrigan.

264 **"You probably just loved that"**: Montville, "On with the Show."

265 **"a nightmare"**: Masters, "Kerrigan Off the Ice Doesn't Seem Half as Nice."

265 **"Overnight, she risked becoming the Shannen Doherty"**: Masters, "Kerrigan Off the Ice Doesn't Seem Half as Nice."

265 **"a semi-celebrity"**: Masters, "Kerrigan Off the Ice Doesn't Seem Half as Nice."

265 **Kerrigan's parents pushed back**: Montville, "On with the Show."

266 **dressed as Harding for Halloween:** Tracy McDowell, interview with the author, July 2015.

266 **"It's amazing she didn't kill anyone":** Matt Harkins, interview with the author, July 2015.

267 **sold to *Penthouse*:** Barry Petchesky, "Woman on Top: Tonya Harding Knew Everything, Says Jeff Gillooly," *Deadspin*, January 16, 2014.

267 **"We do feel a responsibility":** Vivianna Olen, interview with the author, July 2015.

268 **"I remember it was 'Tonya's bad'":** Clara Elser, interview with the author, July 2015.

268 **A Fusion newscast:** Alicia Menendez Tonight and Arielle Castillo, "Yes, These Two Brooklynites Are Really Working on a Museum Devoted to Tonya Harding and Nancy Kerrigan," Fusion, March 11, 2015, http://fusion.net/story/100792/yes-these-two-brooklynites-are-really-working-on-a-museum-devoted-to-tonya-harding-and-nancy-kerrigan/.

268 **Even Barack Obama promised:** Jennifer Parker, "Obama: Not Going to Pull a Tonya Harding," ABC News, December 28, 2007.

269 **"fell, farted, and barfed":** Dan Avery, "'Ice Queens': Watch the Entire Harding-Kerrigan Scandal Now!" Logo, February 24, 2014, http://www.newnownext.com/ice-queens-watch-the-entire-harding-kerrigan-scandal-now/02/2014/.

269 **"still bitter after all these years":** Willa Paskin, "Still Bitter After All These Years," *Slate*, February 24, 2014.

270 **"She didn't really skate that much":** "Nancy Kerrigan's Halloween On Ice Tickets," Ticketmaster, accessed October 2016, http://www.ticketmaster.com/Nancy-Kerrigans-Halloween-On-Ice-tickets/artist/804096?list_view=1.

CHAPTER 12: THE GIRL POWER MYTH

271 **"pretty good at a lot of things":** Greenberg-Lake: The Analysis Group, *Shortchanging Girls, Shortchanging America* (Washington, DC: American Association of University Women, January 1991), https://www.aauw.org/files/2013/02/shortchanging-girls-shortchanging-america-executive-summary.pdf.

272 **"what, on the way to womanhood":** Lyn Mikel Brown and Carol Gilligan, *Meeting at the Crossroads: Women's Psychology and Girls' Development* (Cambridge, MA: Harvard University Press, 1992).

273 **"splits adolescent girls into true and false selves":** Pipher, *Reviving Ophelia*, 35–36.

274 **twenty-six weeks:** Mary Pipher personal website, http://www.marypipher.net/about.html.

276 **"There was nothing that really spoke":** Megan Rosenfeld, "Wholesome Babes in Toyland," *Washington Post*, May 24, 1993.

276 **put up $1 million:** Pleasant Rowland and Julie Sloane, "Chapter 60: How We Got Started—Pleasant Rowland," *Success Story*, last modified July 13, 2004, http://www.angelfire.com/extreme4/success1/ch60.html.

276 **"less fashion and boy-obsessed alternative":** Michelle Leise, "What Are Teen Magazines Telling Your Daughter?; Let's Talk," *Daughters*, September/October 2003.

276 **"affirm self-esteem":** "Our Company," AmericanGirl.com, accessed November 14, 2017, http://www.americangirl.com/shop/ag/our-company.

276 **circulation of over 325,000:** "Our Company," AmericanGirl.com.

277 **"confidence, honesty, innocence, and courage":** Elizabeth Mehren, "Playing with History: In a World of Barbies, What's the Draw of Five Historically Correct Dolls? Confidence and Courage for Starters," *Los Angeles Times*, November 28, 1994, http://articles.latimes.com/1994-11-28/news/ls-2434_1_american-girls-dolls.

277 **$2.7 billion in 1992:** Barbara Brotman, "The Multicultural Playroom," *Chicago Tribune*, October 31, 1993.

277 **$150 million in sales:** Mehren, "Playing with History."

277 **fifteen thousand calls a day:** Mehren, "Playing with History."

277 **That Barbie creator Mattel:** Brotman, "The Multicultural Playroom."

277 **"Totally Hair" Barbie:** "Barbie History—1990s," Mattel, accessed November 8, 2017, http://www.barbiemedia.com/about-barbie/history/1990s.html.

278 **$1 billion a year:** Robert A. Jones, "Barbie Power," *Los Angeles Times*, December 13, 1995.

278 **"choose Barbie over baby dolls":** Rachel Beck, "Some Say No to Barbie, Try to Get Kids Hooked on Other Toys," Associated Press, December 22, 1995.

279 **"becoming a woman on a girl's own terms":** Rose Apodaca Jones, "Cute. Real Cute," *Los Angeles Times*, June 28, 1995.

279 **"It's about labeling":** Joanna Moorhead, "Girl Power Comes of Age," *The Guardian*, October 24, 2007, https://www.theguardian .com/world/2007/oct/24/gender.pop.

280 **By presenting not as women but as girls:** Gayle Wald, "Just a Girl? Rock Music, Feminism, and the Cultural Construction of Female Youth," *Signs* 23, no. 3 (Spring 1998): 585–610, http:// mus15teenpop.weebly.com/uploads/1/6/7/8/1678483/just_a _girl.pdf.

281 **"created a weird intimacy":** Annie Zaleski, "Lisa Loeb Knew She'd Made It When She Heard Herself as Muzak," *VICE*, October 13, 2014, https://noisey.vice.com/en_us/article/6xe77r /lisa-loeb-knew-shed-made-it-when-she-heard-herself-as -muzak.

281 **"Everybody loved to buy into the true story":** Amy Finnerty, interview with the author, January 14, 2016.

282 **"makes it OK to wear":** Maureen Sajbel, "Video Vogue; Singers Finding Some Designer Clothes Suitable," *Los Angeles Times*, July 7, 1994.

282 **"Tortoise-shell cat-eyes":** Joyce Saenz Harris, "Singer Lisa Loeb Takes Her Giant Leap in Stride," *Dallas Morning News*, March 31, 1996.

282 **"something endearing about her out-of-place-ness":** Kim France, "The Last Good Girl," *New York*, September 11, 1995.

282 **"bespectacled Loeb offered":** Margaret Talbot, "Little Women," *New Republic*, January 1, 1996.

283 **"It was so much chatter about her glasses":** Amy Finnerty, interview with the author, January 14, 2016.

283 **"a new cultural dominant":** Wald, "Just a Girl?"

283 **"It's probably because I have glasses":** Lisa Loeb, "17 Questions," *Seventeen*, April 1995.

284 **"I'm very nearsighted"**: Joyce Saenz Harris, "Lisa Loeb; After Instant Fame, She Hopes She's Here to Stay," *Dallas Morning News*, February 18, 1996.

284 **"My glasses are a normal and real part of me"**: Loeb, "17 Questions."

285 **"prove to everybody"**: Kyle Anderson, "Lisa Loeb on Her New Album, the Science of Songwriting, '90s Nostalgia, and the Importance of Desiring Baked Goods," *Entertainment Weekly*, January 28, 2013, http://www.ew.com/article/2013/01/28/lisa -loeb-new-album-no-fairy-tale.

285 **"We remain unconvinced there's much there,"**: Jim Sullivan, "The Year in Rock; The Music Section," *Boston Globe*, December 23, 1994.

285 **"not a great singer"**: Stephen Holden, "Pop Music; Pop Briefs," *New York Times*, November 5, 1995, http://www.nytimes.com /1995/11/05/arts/pop-music-pop-briefs-012190.html.

285 **"girlish wail and tricky lyrics"**: Stephen Holden, "Pop Review; From Out of Nowhere with So Much to Say," *New York Times*, August 29, 1994, http://www.nytimes.com/1994/08/29/arts /pop-review-from-out-of-nowhere-with-so-much-to-say.html.

285 **"I don't know"**: Lisa Loeb, interview with the author, September 2015.

285 **"fail to gain much traction"**: Lisa Loeb, "No Fairy Tale (Bonus Track Version)," iTunes, January 25, 2013, https://itunes.apple .com/us/album/no-fairy-tale-bonus-track/id585902356.

287 **"What are we talking about sex for"**: Steve Pond, "Pop/Jazz; Manufactured in Britain. Now Selling in America," *New York Times*, February 16, 1997.

288 **"I just thought they were the coolest band ever"**: Claire Connors, interview with the author, July 2016.

288 **"Even if this stuff gets out"**: Maria Ricapito, "'Girl' Just Wants to Have Fun," *New York Times*, December 28, 1997.

289 **$75 million per year**: Heidi Sherman, "Ginger Spice's Departure Marks 'End of The Beginning,'" *Rolling Stone*, June 2, 1998, http://www.rollingstone.com/music/news/ginger-spices -departure-marks-end-of-the-beginning-19980602.

289 **Cuddle Core**: Jones, "Cute. Real Cute."

289 **"Love's Baby Soft feminism"**: Ricapito, "'Girl' Just Wants to Have Fun."

290 **"A lot of women can dress"**: Jones, "Cute. Real Cute."

290 **"Us Girls"**: Ricapito, "'Girl' Just Wants to Have Fun."

290 **"being modern, ageless and unfettered"**: Ricapito, "'Girl' Just Wants to Have Fun."

292 **"smatterings of breathlessly excited"**: Jonathan Bernstein, "Get Happy," *Spin*, November 1996.

292 **"petulant whining"**: David Browne, "Tragic Kingdom," *Entertainment Weekly*, August 2, 1996.

292 **"a cross between Jessica Lange"**: Mike Boehm, "The Certainty of No Doubt; Even as It Makes a Splash in the Home Pond, the Anaheim Band Knows to Take Naught for Granted," *Los Angeles Times*, March 16, 1996.

292 **"worthy of the rescue-me blankness"**: Browne, "Tragic Kingdom."

293 **"the paragon of baremidriffed yumminess"**: Bernstein, "Get Happy."

293 **"If it's pretty, she wears it"**: Bob Kurson, "Redoubtable; Success Sweet for 'Girly-Girl,'" *Chicago Sun-Times*, August 9, 1996.

293 **"I love makeup"**: Bernstein, "Get Happy."

293 **Her mother stopped speaking to her**: Jeff Apter, *Gwen Stefani and No Doubt: Simple Kind of Life* (London: Omnibus Press, 2007), 160.

293 **"I don't pay bills"**: Bernstein, "Get Happy."

294 **"I forced Tony"**: Bernstein, "Get Happy."

294 **ten million copies**: "Gold and Platinum: No Doubt, Tragic Kingdom," RIAA, accessed November 11, 2017, http://www.riaa.com/gold-platinum/?tab_active=default-award&ar=No+Doubt&ti=Tragic+Kingdom#search_section.

294 **"sex still sells"**: Browne, "Tragic Kingdom."

294 **"recognized the muscle of girl power"**: Seo, "Magazines for Teens Thrive as Numbers, Buying Power Grow."

295 **Children ages four to twelve commanded**: Laura Liebeck, "Billions at Stake in Growing Kids Market," *DSN Retailing Today* 33, no. 3 (February 7, 1994): 41.

295 **teen girls spent $50 billion**: Seo, "Magazines for Teens Thrive as Numbers, Buying Power Grow."

295 **"You are looking at the future of retail":** "Lisa Frank Captures Girls' Hearts and Parents' Money; Cleaning up in Pro Football Locker Rooms; Yo-Yo Phenomenon has its Ups and Downs," *Business Unusual*, CNN, November 28, 1998.

295 **"It's almost as if they had radar:** "Lisa Frank Captures Girls' Hearts and Parents' Money," *Business Unusual*.

295 **"catering exclusively to the whims":** "Lisa Frank Captures Girls' Hearts and Parents' Money," *Business Unusual*.

296 **the company agreed to pay a $30,000 fine:** Federal Trade Commission, "Web Site Targeting Girls Settles FTC Privacy Charges," news release, October 2, 2001, https://www.ftc.gov/news-events/press-releases/2001/10/web-sitetargeting-girls-settles-ftc-privacy-charges.

296 **the clothing section:** Liebeck, "Billions at Stake in Growing Kids Market."

296 **teen spending reached $122 billion:** Seo, "Magazines for Teens Thrive as Numbers, Buying Power Grow."

296 **claimed that the ad was all about Girl Power:** Michael McCarthy, "Women Say 'No, No, No' to Shampoo Ads; Ad Group Blasts 'Insulting' Spots," *USA Today*, September 27, 2000.

297 **created for girls empowered them:** Patricia Talorico, "Oh Boy, Electronic Fun Is All Dolled Up; Playful Technology Shows Off Softer Side to Get Girls Talking," *USA Today*, February 7, 2000.

297 **92 percent—said they were worried:** Liebeck, "Billions at Stake in Growing Kids Market."

297 **"An ideology based on consumerism":** Amy McClure, "Girl Power Ideology: A Sociological Analysis of Post-Feminist and Individualist Visions for Girls" (presentation, American Sociological Association, San Francisco, CA, August 14, 2004).

297 **"I think they're totally ridiculous":** Karen Schoemer, "The Selling of Girl Power," *Newsweek*, December 29, 1997.

299 **"the pop sensation with an eleventh-grade education":** Steven Daly, "Britney Spears, Teen Queen," *Rolling Stone*, March 29, 2011.

299 **"created as a virgin to be deflowered":** Vanessa Grigoriadis, "The Tragedy of Britney Spears," *Rolling Stone*, February 21, 2008.

299 *The Guinness Book of World Records*: "Angelina, Brad & Britney Set Guinness World Records," *Access Hollywood*, September 15, 2008, http://www.accesshollywood.com/articles /angelina-brad-britney-set-guinness-world-records-65311 /#c7uTBsvXzpDTDvsG.99.

300 **"women now want to be Maxim babes"**: Maureen Dowd, "What's a Modern Girl To Do?" *New York Times Magazine*, October 30, 2005.

302 **"Manolo Blahnik demographic"**: Ashley Fetters, "The New Full-Frontal: Has Pubic Hair in America Gone Extinct?" *Atlantic*, December 13, 2011.

302 **A 2016 study:** T. S. Rowen et al., "Pubic Hair Grooming Prevalence and Motivation Among Women in the United States," *JAMA Dermatology* 152, no. 10 (2016).

EPILOGUE

306 **a total of 104 women:** Jennifer E. Manning and Ida A. Brudnick, "Women in Congress, 1917–2016: Biographical and Committee Assignment Information, and Listings by State and Congress," Congressional Research Service, November 7, 2016, https://fas .org/sgp/crs/misc/RL30261.pdf.

306 **teen pregnancy, births, and abortion have plummeted:** Guttmacher Institute, "U.S. Teen Pregnancy, Birth and Abortion Rates Reach the Lowest Levels in Almost Four Decades," news release, April 5, 2016, https://www.guttmacher.org/news -release/2016/us-teen-pregnancy-birth-and-abortion-rates -reach-lowest-levels-almost-four-decades.

308 **The US maternal mortality rate:** M. F. MacDorman, E. Declercq, H. Cabral, and C. Morton, "Is the United States Maternal Mortality Rate Increasing? Disentangling Trends from Measurement Issues," *Obstetrics and Gynecology* 128, no. 3 (2016).

308 **early weeks postpartum—is rising:** "Maternal Mortality Fell by Almost Half between 1990 and 2015," UNICEF, February 2017, http://data.unicef.org/topic/maternal-health/mater nal-mortality/.

308 **Women were only 17 percent of writers:** Martha M. Lauzen,

"The Celluloid Ceiling: Behind-the-Scenes Employment of Women on the Top 100, 250, and 500 Films of 2016," Center for the Study of Women in Television, 2017, http://womenintvfilm .sdsu.edu/wp-content/uploads/2017/01/2016_Celluloid _Ceiling_Report.pdf.

308 **More women work in television today:** Martha M. Lauzen, "Boxed in 2015–16: Women on Screen and Behind the Scenes in Television," Center for the Study of Women in Television, September 2016, http://womenintvfilm.sdsu.edu/files/2015-16 -Boxed-In-Report.pdf.

308 **Teen births have decreased markedly:** "State Policies on Sex Education in Schools," National Conference of State Legislatures, December 21, 2016, http://www.ncsl.org/research /health/state-policies-on-sex-education-in-schools.aspx.

INDEX

ACKNOWLEDGMENTS

90s Bitch is a social, political, and cultural history, but it was unquestionably shaped by my own reckoning with the misogyny and gender inequality that I saw, felt, and breathed in like air growing up. This book is my attempt to set the record straight, and to rewrite not just the history of women and girls in the 90s, but my own history. I'm indebted to many people for this opportunity and gift.

Thank you Hannah Wood for telling Monika Woods that there should be a book about women in the 90s. I'm grateful to you two for believing in me first, and for encouraging me to run with this. Thank you Richard Pine and Eliza Rothstein for expertly and patiently guiding me through all aspects of the publishing process, and for your high standards, good ideas, and friendship.

Thank you to the crackerjack team of women at Harper Perennial who brought this book into the world, including Stephanie Hitchcock, Amy Baker, Sarah Ried, Megan Looney, Emily Vanderwerken, Trina Hunn, and Suzette Lam. I'm grateful to Becca Giles, Rena Behar, Dori Carlson, Mary Sasso, and Ariel Jicha for their research. I appreciate Mary Beth Constant's copy editing. This book was improved by the expert fact-checking of Matt Savener and Jordan Larson. Thank you thoughtful readers and commenters Frank Flaherty, Corynne Cirilli, Hannah Steiman, Jim Gaudet, and Joby Gaudet. Thank you to all my interview subjects—on and off the record. I'm grateful to those who spoke to me about their experiences in the 90s, and to the women whose stories are shared here in an attempt to reconsider them. Sending gratitude to the many patient and generous editors of my writing, but especially Harry Siegel and Gabrielle Birkner.

Big-ups to my TEDRes 4 fam—Alvin, Anouk, Bob, Derrius, Eiji, Jason, Karen, Kifah, Malika, Michael, Tobacco, Stan, Will, Cyndi, and Katrina. You kept me inspired, challenged, hydrated, and in good company.

Ruby was born a week before the sale of *90s Bitch* was announced, and Oscar arrived on the cusp of the first round of edits. I admire and appreciate Arlene Fender, and thank her deeply for loving and caring for our kids so that my husband, Ben, and I can work.

Leah and Bill Yarrow are fantastic and writerly in-laws for whom I am grateful. You always get me talking and thinking. Thank you Jami and Jim Gaudet for your unyielding support, love, and curiosity. You have no small role in the fact that I think and see the world the way that I do. Joby Gaudet, you amaze me and make me stupidly proud. I'm grateful for you in too many ways to count here.

Ruby and Oscar, I love you infinity. You aren't old enough to read, and you probably shouldn't say the title of the book out loud in public, but I'll leave this here for you for later—I believe in each of you and will always fiercely support you in exactly who you are, and who you want to become. I hope things that were hard for me won't be hard for you, that the world has changed since the 90s, and that it will continue to change and become more loving and just. Let's keep working on it.

I heard someone say that family members make the best editors because they can speak blunt truth to you. Maybe, but that idea is complicated when you're married to someone who also happens to be the best editor you've ever encountered. I'm lucky but also, well, shit. Ben, had I known sooner, I might have married you on our first date. This book wouldn't be what it is without you. You thrill and inspire me, and you're the best man I've ever met that I'm not related to by birth. Thank you for your love, support, and for our life.

ABOUT THE AUTHOR

Allison Yarrow is an award-winning journalist and National Magazine Award finalist whose work has appeared in the *New York Times*, *Washington Post*, *Vox*, and many others. She was a TED resident and is a grantee of the International Women's Media Foundation. She produced the Vice documentary *Misconception*, and has appeared on the *Today* show, MSNBC, NPR, and more. Allison was raised in Macon, Georgia, and lives in Brooklyn, New York.